SpringerBriefs in Education

We are delighted to announce SpringerBriefs in Education, an innovative product type that combines elements of both journals and books. Briefs present concise summaries of cutting-edge research and practical applications in education. Featuring compact volumes of 50 to 125 pages, the SpringerBriefs in Education allow authors to present their ideas and readers to absorb them with a minimal time investment. Briefs are published as part of Springer's eBook Collection. In addition, Briefs are available for individual print and electronic purchase.

SpringerBriefs in Education cover a broad range of educational fields such as: Science Education, Higher Education, Educational Psychology, Assessment & Evaluation, Language Education, Mathematics Education, Educational Technology, Medical Education and Educational Policy.

SpringerBriefs typically offer an outlet for:

- An introduction to a (sub)field in education summarizing and giving an overview of theories, issues, core concepts and/or key literature in a particular field
- A timely report of state-of-the art analytical techniques and instruments in the field of educational research
- A presentation of core educational concepts
- An overview of a testing and evaluation method
- A snapshot of a hot or emerging topic or policy change
- An in-depth case study
- A literature review
- A report/review study of a survey
- An elaborated thesis

Both solicited and unsolicited manuscripts are considered for publication in the SpringerBriefs in Education series. Potential authors are warmly invited to complete and submit the Briefs Author Proposal form. All projects will be submitted to editorial review by editorial advisors.

SpringerBriefs are characterized by expedited production schedules with the aim for publication 8 to 12 weeks after acceptance and fast, global electronic dissemination through our online platform SpringerLink. The standard concise author contracts guarantee that:

- an individual ISBN is assigned to each manuscript
- each manuscript is copyrighted in the name of the author
- the author retains the right to post the pre-publication version on his/her website or that of his/her institution

Tengteng Zhuang

Modernizing China's Undergraduate Engineering Education Through Systemic Reforms

Ideas, Practices, and Impacts

 Springer

Tengteng Zhuang
Institute of Higher Education
Faculty of Education
Beijing Normal University
Beijing, China

ISSN 2211-1921   ISSN 2211-193X (electronic)
SpringerBriefs in Education
ISBN 978-981-99-6387-4   ISBN 978-981-99-6388-1 (eBook)
https://doi.org/10.1007/978-981-99-6388-1

This Springer imprint is published by the registered company Springer Nature Singapore Pte Ltd.
The registered company address is: 152 Beach Road, #21-01/04 Gateway East, Singapore 189721, Singapore

Paper in this product is recyclable.

Contents

About the Author

Dr. Tengteng Zhuang is currently an assistant professor/lecturer at Institute of Higher Education, Faculty of Education, Beijing Normal University, which is one of the leading institutions in education science in Asia and the world at large. Having a dual background of engineering and education, he is focused on various issues of higher education, such as teaching effectiveness, university student development, and university-industry collaboration, particularly in relation to the field of engineering. He has published peer-reviewed academic articles covering topics of engineering students' program satisfaction, engineering faculty members' teaching agency, engagement, determinants and challenges to teaching-focused university-industry collaboration, and other higher education-related issues. His publications have appeared in various SSCI journals, such as *British Journal of Sociology of Education, Higher Education Research and Development, Cambridge Journal of Education*, and *Asia Pacific Education Review*. Furthermore, he has frontline work experiences as an insider of the Chinese higher education sector, with experiences of both teaching and administrative work. He has been the principal investigator of several projects pertaining to teaching-focused university-industry collaboration in China and Asia in recent years.

About the Author

Chapter 1
Introduction

1.1 Research Background and Purposes

Globally, engineering education is a highly important branch of higher education. In many countries, the majority of higher education is occupied by engineering majors. The level and quality of engineering in a country are closely related to its economic development and overall national strength. The quality of engineering education is considered crucial in determining a country's overall engineering level. Engineering education is responsible for training high-quality and high-level engineering talents. In the official documents of many governments and universities, the significance of engineering education is not limited to the ivory tower. The importance and far-reaching influence of engineering education are always related to the long-term destiny of society and the country.

Almost every country in the world attaches importance to the development of engineering education. As a discipline, engineering education has a long history dating back to the United States in the late nineteenth century. For instance, in America, the Society for Promotion of Engineering Education (SPEE) was established as early as 1893. It marked the beginning of the systematic study of teaching methods and curriculum development in engineering education as a distinct field (Lohmann & Froyd, 2010). Subsequently, over the past century, different countries worldwide have established engineering education organizations under different names. In 1989, the International Engineering Alliance (IEA) was formally established through the signing of the Washington Accord by six founding members: Australia, Canada, Ireland, New Zealand, the United Kingdom, and the United States. It marked the beginning of the exploration of engineering education together by different countries or regions through cooperation, interaction, and grouping. They also recognized each other's engineering education achievements. According to the official documents of IEA, the international engineering alliance is concerned with engineering, education, and competence across the entire spectrum of engineering. Its core activities include consistent improvement of standards and mobility, defining standards of

© The Author(s), under exclusive license to Springer Nature Singapore Pte Ltd. 2023
T. Zhuang, *Modernizing China's Undergraduate Engineering Education
Through Systemic Reforms*, SpringerBriefs in Education,
https://doi.org/10.1007/978-981-99-6388-1_1

education and professional competence, assessment of education accreditation and evaluation of competence, and participation in activities driven by the engineering profession (International Engineering Alliance, 2014). The Washington Accord is a constituent of the International Engineering Alliance, consisting of educational accords for professional engineers, engineering technologies, engineering technicians, and some professional engineering competency agreements. It has provided a mechanism for mutual recognition of graduates of accredited programs among its signatories. Although the Washington Accord had only six member countries at the beginning of its establishment, it has rapidly grown over the following 30 years. Today, the signatories of the Washington Accord include many industrially advanced countries and regions in the world, such as Hong Kong (China), South Africa, Japan, Singapore, Korea, Chinese Taipei, Malaysia, Turkey, and Russia. Several other countries, such as Bangladesh, China, India, Pakistan, the Philippines, and Sri Lanka, have gradually joined the signatories after applying and holding provisional status in recent years. In fact, the agreements under the International Engineering Alliance include not only the Washington Accord but also other agreements related to engineering education, such as the Sydney Accord and the Dublin Accord. These agreements indicate that international engineering education focuses on cultivating engineering talents in various aspects, which are critical in fueling economic development, ensuring national and global security, and achieving sustainable development goals by addressing contemporary challenges.

For example, the Washington Accord stipulates 12 aspects of graduate attributes for engineering students to achieve, including engineering knowledge, problem analysis, design or development of solutions, investigation, modern tool usage, the engineer and society, environment and sustainability, ethics, individual and teamwork, communication, project management and finance, and lifelong learning. With respect to engineering knowledge alone, it identifies a total of eight important elements. It not only requires students to have a systematic theory-based understanding of the natural sciences applicable to the discipline but also expects engineering graduates to master conceptually-based mathematics, numerical analysis, statistics, and formal aspects of computer and Information science to support analysis and modeling applicable to the discipline. At the same time, it also wants students to have a good mastery of engineering specialist knowledge that can provide theoretical frameworks and bodies of knowledge for the accepted practice areas in the engineering discipline. Furthermore, it requires students to master the skills to resolve complex engineering problems that have a range of attributes. For example, students are expected to gain depth of knowledge, range of conflicting requirements, depth of analysis, familiarity of issues, extent of applicable codes, extent of stakeholder involvement and needs as well as interdependence in their learning. And to facilitate the achievement of these goals, the Washington Accord maintains that some resources must be in place to ensure that students and instructors have external support to achieve intended outcomes.

Other engineering education agreements, such as the Sydney Accord, Dublin Accord, and APEC Agreement, also have different stipulations and requirements for the cultivation of engineering talent. High-quality engineering education is a systematic project that involves many internal and external factors and conditions, such as students, teachers, schools, systems, environmental policies, and so on if intended outcomes are to be achieved.

Engineering education has always occupied a very important position in China's higher education landscape in China's modern history. Since the Opium War, especially after the late 1890s, the overall national strength gap between China and Western industrialized countries has repeatedly proven that the essence lies in the gap in science, technology, and industrialization advancement, which led to China's repeated defeat and suffering in wars in the subsequent 'century of humiliation'. Compared with feudal dynasties, engineering education began to receive unprecedented attention in the era of the Republic of China. The government and various sectors invested in and supported engineering education in an unprecedented manner. Since the founding of the People's Republic of China, the whole country has started from scratch after the civil war. Vigorously developing the level of industrialization and transforming the country from an agricultural country to an industrialized power has been an unswerving pursuit of Chinese governments at all levels for more than half a century. Whenever there are some great achievements in the field of science and engineering in modern China, the Chinese people will be very proud and widely publicize them. Whether it's the "two bombs and one satellite" in times of national poverty or high-speed railways or the mega Hong Kong-Zhuhai-Macau Bridge in contemporary wealthy times, great achievements in various engineering fields can always be successfully publicized in the spotlight. All this progress and development cannot be separated from the backup support of millions of engineering talents.

Among China's colleges and universities, a large proportion of top-class institutions are engineering-focused, such as Tsinghua University, Shanghai Jiaotong University, Harbin Institute of Technology, Xi'an Jiaotong University, Northwestern Polytechnical University, Beihang University, and University of Chinese Academy of Sciences. These institutions have gradually gained recognition from international counterparts, as evidenced by their positions in various world university rankings. Since the beginning of the twenty-first century, China's Ministry of Education has often emphasized the importance of engineering education in higher education, using the words "half of all" to describe its significance. According to the 2013 China Engineering Education Quality Report, there were 1077 undergraduate colleges and universities offering engineering majors, accounting for more than 90% of the total number of undergraduate colleges and universities in China. There were 4,953,300 engineering undergraduate students, representing about 1/3 of the total number of undergraduate students in colleges and universities in China. Additionally, engineering programs occupy about 1/3 of all programs in the Chinese higher education sector. In terms of scale, China's university-level engineering education is the largest in the world, and the Ministry of Education often calls China a leader in engineering education globally.

Despite these achievements, academia and the government of China are aware that there are still challenges to be overcome in China's university-level engineering education. Although China has become a major player in engineering education, it is still far from a powerful country in this field. The application ability and practical skills of engineering graduates to solve complex engineering problems are still areas of concern in China's engineering education. This challenge is particularly signifi- cant in the global context of Industry 4.0, where scientific and technological advances are accelerating in the industry sector. New technologies, such as big data automa- tion, artificial intelligence, the Internet of Things, cloud computing, biomedical 3D printing, and unmanned driving, are constantly evolving and developing, creating a demand for high-level engineering talents in both quality and quantity. Therefore, reforming engineering education and improving the quality and efficiency of training engineering talents is a priority for China's higher education sector.

Since the beginning of the twenty-first century, especially in the last decade, the pace of engineering education reform in China has accelerated significantly. In 2016, China became a member of the Washington Accord, and the Ministry of Higher Education has strongly emphasized the concept of international substantive equivalence in developing engineering education since then. Many top-level designs, policies, and reform models related to engineering education have been launched in recent years.

This book will focus on the reform policies, practices, and impacts of engineering education in China in recent years. It aims to comprehensively sort out the prac- tices, experiences, and challenges involved in the course of engineering education reform in the new era. This book is suitable for academic researchers in higher educa- tion reform, particularly those interested in engineering education, higher education institution leaders, educational officials in government authorities, and anyone inter- ested in engineering education at the university level. The book will present not only policies, discourse, or intentions at the macro level, but also micro-level practices, examples, and cases. Additionally, the book will not only present the current state of engineering education in China but also scrutinize the impact of relevant reform agendas.

1.2   Distinctive Features of Engineering Disciplines and Requirements on Engineering Education

Every discipline has unique features that other disciplines do not possess, and engi- neering disciplines are no exception. Many studies have shown that the attributes and characteristics of engineering disciplines are significantly different from those of humanities and social sciences. This difference can be viewed from two aspects: theory and practice.

Theoretically, many engineering disciplines are fundamentally based on complex, abstract, and advanced mathematics. General theorems in many engineering practices have clear mathematical expressions as the underlying basis. If one's mathematical foundation is weak, it becomes difficult to comprehend some engineering principles even after learning them to a certain level. Whether it is communication engineering, optical engineering, or automatic control engineering, no excellent engineer or engineering talent can have a poor mathematical foundation. Engineering undergraduates have to study many mathematics-related courses in the first or even second year, such as introductory advanced mathematics, linear algebra, probability theory, and more professional courses like engineering mathematics and physical mathematics. These courses often involve the bottom-level operating rules in the field of science and engineering.

In addition to mathematics subjects, each engineering field has its own professional theoretical foundation courses. For example, the professional basic courses in communication engineering include communication principles, digital signal processing, signal and system, analog electronic communication principles, and digital electronic communication principles. The theoretical foundational courses of software engineering include programming language, computer application foundation, software technology foundation, algorithm design, operating systems, data structure, and many others. The abstraction and obscurity of these courses are much greater than those in high school subjects. On the one hand, such abstraction is reflected in the significant increase of obscure concepts and technical terms encountered during professional learning. On the other hand, it is more reflected in the fact that such concepts often cannot find their illustrative and visible counterparts in the learners' real life. For example, the quadratic function, which is somewhat challenging in basic education, can still be easily understood by students through the opening direction of the parabola, the positions of the symmetry axis and origin, and the amplitude of the parabola. However, at the higher education level, the basic theory of many engineering principles, such as Fourier series, aimed at expressing complex periodic functions into various linear combinations of trigonometric functions, not only becomes much more complex, but also more difficult to find a direct corresponding mapping in real life. Fourier series is the basic theory in many engineering disciplines, such as communication engineering, electronic information technology, and automatic control. According to the requirements of the International Engineering Certification Association, a mature engineering program must teach students enough theoretical knowledge through its curriculum system, but the deeper the knowledge, the more abstract the theoretical basis used. This indicates that the obscurity and abstractness of the theory of science and engineering education majors in colleges and universities pose tremendous cognitive challenges to learners. Some scholars have even used the phrase "abstract shock" to aptly describe the discomfort experienced by high school students after entering university as they cannot adapt to the learning of theoretical knowledge (Griese, 2017).

If theory is the fundamental characteristic of any discipline, then practical application makes engineering disciplines stand out. Knowing only theory but not how to apply it is not what engineering is all about. Numerous studies demonstrate the differences between engineering as a hard discipline and other soft disciplines, as well as the distinct requirements for students' application competence, such as the differences between engineering and business (Chan & Fong, 2018) and between engineering and humanities and social sciences (Zhuang et al., 2019). Engineering graduates are expected to demonstrate the ability to apply knowledge of math, science, and engineering fundamentals to real-life scenarios, design and conduct experiments, analyze and interpret data, design systems, components, or processes to meet desired social needs, function proficiently on multi-disciplinary teams, and more (CSU, 2016). The Washington Accord mentioned above also makes specific requirements for the application ability of engineering graduates, stating that they should be able to "design solutions for complex engineering problems and design systems, components or processes that meet specified needs with appropriate consideration for public health, and safety, cultural, societal and environmental considerations" and possess the ability to "create, select and apply appropriate techniques, resources and modern engineering and IT tools, including prediction and modelling, to complex engineering problems, with an understanding of the limitations" (International Engineering Alliance, 2014). However, in reality, the disconnection between academia and industry is a long-standing issue that has not been well resolved in the engineering sector and is perplexing university administrators, industrial employers, government officials, and other stakeholders of higher education across countries (National Academy of Engineering, 2005; Zhuang & Zhou, 2022).

Moreover, engineering disciplines are often highly interdisciplinary, which is particularly the case in the context of Industry 4.0. On the one hand, in the industry, there are almost no engineering projects or products that are completed by knowledge or skills from a single discipline. The conception, design, implementation, and operation of any specific product require close cooperation among personnel from different departments and disciplines. On the other hand, in the reform of higher engineering majors on campus, more and more emphasis is placed on breaking down the barriers between disciplines, leading teachers of different majors and disciplines to start classes together for students. While the reform of modern engineering education should emphasize the wholeness and integrity of knowledge and thinking in professional fields, realizing stronger interdisciplinarity is also one of the important directions of current engineering education reforms (Graham, 2018). Engineering instructors are increasingly expected to present students with numerous interdisciplinary cases and real engineering project cases in their teaching process. Engineering education needs to break through the current definition of disciplines and specialties and embody the integration of multidisciplinary thinking, new industrial technology and engineering discipline theory, and the integration of students' interdisciplinary professional ability. Teachers are urged to select, organize, and integrate curriculum contents from different disciplines to meet the needs of future-oriented industry, professional development, and personnel training, effectively supporting the realization of graduate attributes. Desired practices include linking and coordinating traditional single discipline-based courses with multiple disciplined curriculum resources to support and reinforce the interdisciplinary aspects of engineering methods and problem-solving. Existing research also shows that chemistry studies can greatly promote college students' knowledge construction and achievement improvement, and at the same time, affect students' learning psychology, learning behavior, and teaching satisfaction.

The unique attributes of engineering disciplines also put forward many unique requirements for the practice of engineering teaching in the classroom. The Conception, Design, Implementation, and Operation (CDIO) engineering education model is the guiding concept for the development of science and engineering majors in hundreds of renowned universities worldwide. Among its 12 recognized teaching standards of engineering education, many standards clearly reflect the significant influence of instructors' classroom teaching methods.

For example, the 10th standard emphasizes the improvement of teachers' teaching abilities, indicating that integrated teaching abilities should be improved at the undergraduate level on a global scale, and students' learning should be promoted by actively innovating teaching methods. The 3rd standard, which aims to create a comprehensive curriculum, clearly points out that the curriculum system of science and engineering majors should be based on a modular training scheme in which different but related disciplines support each other. It should also focus on combining students' knowledge and skills with multidisciplinary courses, and teachers should actively establish interdisciplinary links when designing courses. The 6th standard calls for the use of workspaces and laboratories to support and encourage hands-on learning of product and system building, disciplinary knowledge, and social learning.

1.3 The Structure of the Book

This book focuses on engineering education at the university level in China and the various reforms that have taken place in recent years. Its structure consists of seven chapters.

Chapter 2 introduces the New Engineering Education (NEE), which is the most significant reform project in China's higher education sector since 2017. The rationale behind this chapter is that NEE serves as a comprehensive framework that encompasses other specific reform measures. It acts as an umbrella, guiding the implementation of these measures. Within this chapter, we will first present the trilogy policy documents of NEE, namely the Fudan Consensus, Tianda Action, and Beijing Guide. These policy documents outline the plans and assumptions underlying different aspects of engineering education reform, embodying the overall vision of China's higher engineering education reform. After discussing the national-level policies, this chapter will also explore the local and institutional-level reform plans, which align with the national policies but also incorporate region-specific elements.

Chapter 3 focuses on another crucial aspect of engineering education reform: the implementation of industry-university cooperation in cultivating engineering talent within the Chinese context. This chapter addresses several key issues, including the recognition within Chinese higher education of the importance of collaboration between enterprises and universities for teaching purposes. It also explores the role of the government in coordinating the interests of schools and enterprises, the process of selecting suitable partners between universities and enterprises, the collaborative efforts of all stakeholders in establishing a higher education innovation ecosystem supported by appropriate incentive mechanisms and resources, and the specific measures employed by the industry to enhance the quality of education and teaching in higher education institutions.

Chapter 4 delves into the impact of engineering education reform on classroom education and teaching at the micro-level. The ultimate goals of these reforms are centered around education, teaching, and students' learning. Therefore, if there are no substantial changes in education and teaching, all reform efforts remain indirect and peripheral. This chapter explores how reforms, in the context of NEE initiatives, aim to visualize and contextualize complex theoretical knowledge. It also examines the collaborative virtual teaching and research communities established by numerous universities in China in recent years. Furthermore, the chapter highlights China's achievements and experiences in utilizing Internet technology to implement large-scale massive open online courses (MOOCs) over the past decade, with the aim of disseminating high-quality course content to benefit a wider student population.

Chapter 5 centers on the role of discipline competitions and student contests in engineering education and their contributions to student learning. Compared to traditional classroom learning, these scientific research competitions and subject competitions are regarded as the "second classroom" in higher education and have received increased emphasis in various policy documents in recent years. The chapter provides readers with a selection of diverse student competitions, showcasing practical avenues for engineering students to strive for excellence and enhance their learning efficiency beyond the confines of the traditional classroom.

Chapter 6 focuses on two notable practices in the process of engineering education reform: the establishment of modern industrial colleges and future technical colleges. These approaches represent innovative frameworks within engineering education, designed to align with the latest practices in contemporary industrial needs and future technological development trends. The chapter introduces the seven key tasks outlined by the Ministry of Education for establishing modern industrial colleges and provides a list of the initial 50 modern industrial colleges selected nationwide by the Ministry of Education. Additionally, it explores crucial aspects related to the establishment and enhancement of future technical colleges.

Chapter 7 focuses on China's endeavors in engineering education research. This chapter presents 17 key aspects of engineering education research as strategic measures that require attention and implementation. Particularly noteworthy is the formal inclusion of engineering pedagogy as a distinct discipline, a development that significantly enhances the disciplinary status of engineering education research within the broader landscape of higher education research.

Chapter 8 serves as the concluding chapter of this book, providing a comprehensive summary and assessment of the initial impact of various reform measures in higher engineering education in China. It examines the practical effects resulting from these reform measures, considering both structural and cultural aspects. Additionally, this chapter analyzes the challenges faced by engineering education reform and offers insights into future prospects within the field.

Acknowledgements This work was supported by Beijing Normal University First-class Discipline Cultivation Project for Educational Science (Grant Number: YLXKPY-ZYSB202212). The Project Name is "Patterns for the Industry Sector to Boost Higher Education Quality Development in Technology-savvy Asian Countries under the New Round of Industrial Revolution".

References

Chan, C. K., & Fong, E. T. (2018). Disciplinary differences and implications for the development of generic skills: A study of engineering and business students' perceptions of generic skills. *European Journal of Engineering Education, 43*(6), 927–949.

CSU. (2016). *Program improvement process: Electrical/Electronic engineering* (4th edn). http://www.csuchico.edu/eece/_assets/documents/CMPE_Program_Improvement_Process_S16.pdf

Graham, R. (2018). *The global state of the art in engineering education.* Massachusetts Institute of Technology

Griese, B. (2017). *Learning strategies in engineering mathematics.* Springer Fachmedien Wiesbaden. https://doi.org/10.1007/978-3-658-17619-8

International Engineering Alliance. (2014). *25 years of the Washington Accord.*

Lohmann, J., & Froyd, F. (2010). Chronological and ontological development of engineering education as a field of scientific inquiry. In *Second Meeting of the Committee on the Status, Contributions, and Future Directions of Discipline-Based Education Research*, Washington, DC. Available: http://www7.nationalacademies.org/bose/DBER_Lohmann_Froyd_October_Paper.pdf

National Academy of Engineering. (2005). *Educating the engineer of 2020: Adapting engineering education to the new century.* National Academic Press.

Zhuang, T., & Zhou, H. (2022). Developing a synergistic approach to engineering education: China's national policies on university–industry educational collaboration. *Asia Pacific Education Review*, 1–21.

Zhuang, T., Cheung, A. C. K., Lau, W. W. F., & Tang, Y. (2019). Development and validation of an instrument to measure STEM undergraduate students' comprehensive educational process. *Frontiers of Education in China, 14*(4), 575–611.

Chapter 2
The Umbrella Reform Initiative in the New Times: New Engineering Education

Abstract This chapter provides a comprehensive overview of the paramount design documents related to the recent reforms in China's higher engineering education, the New Engineering Education (NEE) paradigm, and the associated policies. Not only does it elucidate the distinct aspects emphasized by China's engineering education reform, but it also details how several prominent regional institutions have crafted localized and specialized strategy documents in alignment with this paramount design. Included among these are the strategic approaches of Tianjin University, University of Electronic Science and Technology, and South China University of Technology. Through this introduction to the paramount document, readers can gain a foundational understanding of the overall reform initiatives undertaken in China's engineering education at university level in recent years.

Keywords New engineering education · Engineering education reform · China's higher education

2.1 The Trilogy of the National "New Engineering Education (NEE)" Policies

In the previous chapter, we mentioned a very important mutual recognition agreement for engineering education under the International Engineering Alliance, the Washington Accord. The Washington Accord is an agreement on the mutual recognition of undergraduate professional degrees in engineering education, with the purpose of promoting mutual recognition of engineering degrees and international mobility of engineering and technical personnel through multilateral recognition of engineering education qualifications. The main contents of the Washington Accord include the following: the engineering professional certification standards, policies, and procedures adopted by all formal signatories are basically equivalent; all full members recognize each other's authentication results provided by other full members and issue statements in an appropriate manner to recognize the results; all formal signatories should promote professional engineering education to realize the educational

© The Author(s), under exclusive license to Springer Nature Singapore Pte Ltd. 2023   11
T. Zhuang, *Modernizing China's Undergraduate Engineering Education Through Systemic Reforms*, SpringerBriefs in Education,
https://doi.org/10.1007/978-981-99-6388-1_2

preparation needed for engineering professional practice; and all full members should maintain mutual supervision and information exchange. In fact, the Washington Accord is also an important driving force for China to launch a new round of engineering education reform.

Aware of the importance of obtaining international substantive equivalence and professional recognition from international peers, since 2006, China has begun to make systematic and serious preparations for joining the Washington Accord (Zhuang & Xu, 2018). In 2012, China formally submitted its application to join the Washington Accord, and after 10 years of hard work, China became a formal signatory in 2016. Joining the Washington Accord is a significant event in the field of higher education and higher engineering education in China, and this important opportunity has been shared and reprinted by major domestic media. For Chinese higher education circles, joining the Washington Accord means that the quality of engineering majors that China has worked hard to build for many years has been recognized by international counterparts. Therefore, after joining the Washington Accord, one of the most discussed topics and key words among academic circles and government officials is substantive equivalence.

After joining the Washington Accord, China aims to further adjust and optimize the existing engineering specialty setting and the quality of engineering education in accordance with the relevant requirements of the Washington Accord. Under this specific educational background, the Ministry of Education of China officially launched the state-level engineering initiative, the NEE, in 2017 (Koh & Zhuang, 2021). According to Yan Wu, the then director of the Department of Higher Education of the Ministry of Education, the development of NEE in China is an educational response to the impact of global technological and industrial revolution on higher education. The initial advancement of NEE is marked by the issuing of three series policy documents: Fudan Consensus, Tianda Action, and Beijing Compass (Koh & Zhuang, 2021). The three policy documents have made corresponding arrangements for different aspects of the development of NEE.

The Fudan Consensus begins by describing a technologically changing world driven by innovation. For instance, it clearly states that "a new round of technological revolution and industry upgrade is accelerating its pace worldwide" [Paragraph 1 of Fudan Consensus, cited in Koh and Zhuang (2021)] and warns that those who can't keep pace with the times and keep up with the frontier of the times will be eliminated by the cruel global competition. However, the essential function of the Fudan Consensus does not lie in setting the harsh global technological competition but is more about defining the roles that different stakeholders in higher engineering education should play, including the different responsibilities of engineering-savvy universities, comprehensive universities, local universities, government departments, and social forces.

The Fudan Consensus requires universities with engineering advantages to play a main role in engineering science and technology innovation and industrial innovation (MOE, 2017a). These colleges and universities have high hopes. They should sum up and inherit the successful experience of engineering education reform and development, deepen the reform of engineering and personnel training, give full play to their advantages of close ties with industries, and actively optimize the layout of disciplines and specialties for the current and future industrial development. In ensuring a higher degree of interdisciplinary engineering education, universities with engineering advantages are required to promote the cross-review of existing engineering, the cross-integration of engineering and other disciplines, actively develop new engineering, expand the connotation and construction focus of engineering majors, build an innovative value chain, and vigorously cultivate engineering science and technology innovation and industrial innovation talents to serve, upgrade, and transform industries.

As for the responsibilities of comprehensive universities, the Fudan Consensus requires such universities to give full play to their comprehensive advantages in disciplines and play a leading role in promoting new technologies and breeding new industries. They are also expected to promote interdisciplinary integration and boundary-spanning integration to generate new technologies, cultivate new engineering fields, promote the organic integration of science education, humanities education, and engineering education, and then cultivate talents with strong engineering ability and comprehensive engineering competence.

Regarding local colleges and universities, the Fudan Consensus asks these institutions to play a supporting role in regional economic development and industrial transformation and upgrading. Local colleges and universities should take the initiative to meet the needs of local economic and social development and technological innovation of enterprises, make full use of local resources to give full play to their advantages, deepen the integration of production and education, school-enterprise cooperation, collaborative education, enhance students' employment and entrepreneurship ability, and cultivate a large number of applied, technical, and skilled talents with industry background knowledge, engineering practice ability, and competence for the development needs of the industry.

In addition, the Fudan Consensus also clearly points out that the construction of NEE needs strong support from government departments. According to this document, the Ministry of Education, relevant industry authorities, and governments at all levels should focus on supporting the construction of new engineering education, promoting the reform of system and mechanism, strengthening policy coordination, and introducing more support measures in optimizing the training mechanism of professional structure reform in related fields, strengthening practical training, strengthening the construction of teaching staff, etc., to provide a good policy environment for the training of new engineering education talents. The Fudan Consensus also discusses the ushering in of social forces to participate in the carrying out of NEE, as well as strengthening international cooperation and learning from international experience in the last part.

The second document, Tianda Action, focuses on clarifying six aspects that need to be reformed in the process of promoting NEE. Interestingly, this document employs the use of juxtaposition of six subheads to indicate the areas of prospective reforms, in the language pattern of 'studying + (an area) + to (work on) + (an aspect), in order to (achieve a goal)' (MOE, 2017b). Specifically the six subjects go as:

"Studying demands of industry to establish and develop programs, in order to institute a new structure for engineering programs";

"Studying technological development to change contents, in order to upgrade the system for cultivating engineering talents";

"Studying students' interest to change methods, in order to innovate teaching approaches and measures for engineering education";

"Studying institutions' accountability to promote reform, in order to explore mechanism for self-development and self-motivation";

"Studying resources available to create conditions, in order to build an open and inclusive engineering education"; and

"Studying international frontiers to set standards, in order to enhance engineering education's global competitiveness".

It can be seen that the six major reform aspects emphasized by Tianda Action include upgrading structures of engineering specialties, renewing the knowledge system of engineering talents, innovating engineering education methods and means, exploring a new incentive mechanism for engineering development, creating a new ecology of engineering education, and enhancing the international competitiveness of engineering education. While identifying six important reform areas, Tianda Action also highlights some emerging engineering fields that need the attention of China's higher engineering education, including big data, cloud computing, Internet of Things applications, artificial intelligence, virtual reality, genetic engineering, nuclear technology and intelligent manufacturing, integrated circuits, aerospace and ocean, biomedicine, and new materials. For some traditional engineering majors such as geology, mineral resources, petrochemical, iron and steel, textile machinery, etc., it is required that these majors should constantly upgrade and update the corresponding contents and be as close to the cutting-edge technologies in related fields as possible.

The third document, 'Beijing Compass,' mainly pertains to the five paths which NEE reform should follow, including conceptual update, structural optimization, pattern innovation, quality assurance, and categorical development (MOE, 2017c). For each path of the five, existing problems in the current undergraduate engineering education are identified. For example, regarding the area of pattern innovation, it states that "the barriers that prevent extensive social participation in the development of engineering education must be overcome by improving a mechanism of multi-agent collaborative education based on closer bonds between scientific development and engineering education, between industrial advancement and higher learning, and between universities and enterprises" [Beijing Compass, cited through Zhuang and Xu (2018)]. In addition, Beijing Compass also visualizes a series of achievements that NEE is supposed to achieve eventually. These outcomes include a multiplicity of high-quality universities of science and technology, a multiplicity of colleges of industrialization co-developed and co-managed by multiple stakeholders, a multiplicity of new polytechnic programs that meet industry demands, a multiplicity of new curricula that reflect the state-of-the-art technology, a multiplicity of praxis platforms that integrate education, training and R&D, a multiplicity of highly professional instructors with strong competence in engineering praxis, a multiplicity of cross-disciplinary platforms for the R&D of new technologies, a multiplicity of local industry-oriented platforms for technological innovation, and a multiplicity of other transferable reform outcomes (Zhuang & Xu, 2018).

In fact, each paragraph of the three series of documents of NEE policy contains a relatively broad range of fields on which reform outcomes should be achieved. Due to the space limit, they cannot all be listed here. In the words of Wu Yan, Deputy Minister of Education of the People's Republic of China, the 'new' elements of NEE are mainly supposed to be reflected in three aspects: the new engineering programs, new requirements of engineering programs and specialties, and interdisciplinarity of engineering programs. At the same time, Wu Yan argues that it is not enough for NEE to bring changes only at the micro level such as curriculum, teaching materials, and professional teaching. He believes that the new engineering education also needs to go through a surgery-like reform in terms of structural conditionings out of the perception that a sector's function is often determined by the structural arrangements and configurations of its constituents. Therefore, in a public speech at Peking University, he pointed out that the future direction of new engineering construction also includes the establishment of a large number of future technical colleges and modern industrial colleges and characteristic professional colleges, so as to ensure that the whole engineering education is structured to be more student-centered and learner-friendly.

Wu further argues that the role of NEE is not only to promote the reform of education and teaching in related fields of engineering specialty but also to promote the overall reform of the whole higher education pattern in China. Whereas the first mission of NEE is to cultivate a large number of extraordinary quasi-engineers for the country by concurrently resolving the issues in textbook development, instructor professionalism, system integration, and training basis construction, NEE is also anticipated to "lead the reform and innovation of higher education and empower the future," as Wu Yan said in one of his public speeches in Shanghai in 2019.

In a later article, Wu summarizes that NEE development should focus on five aspects: advancing theories concerning engineering education, developing new programs to cater to real-life demands, changing structural configurations to cultivate talents, upgrading courses and contents for teaching, and facilitating boundary-spanning integration. In the aspect of theoretical advancement, two batches of new engineering research and practice projects are meant to promote the academic circles to renew their ideas on the concept of talent training in engineering education. Regarding establishing specialties, the goal of promoting new engineering education is to push forward the shortage strategy and new specialties that society needs more in a down-to-earth manner. Wu says that more than 1/4 of all engineering majors are new majors in short supply of national strategy. Therefore, in the overall adjustment process, the Ministry of Education canceled 251 engineering majors across Chinese universities in 2021, but 794 new engineering majors were added on the basis of "Internet+" and "Smart+".

In terms of changing the structure, some work done in recent years has a time span. For example, Modern Industrial College aims at the industrial changes after five years and trains high-end applied talents and high-quality business talents. Individualized software colleges, demonstration microelectronics colleges, etc., aim at the talents needed by the industrial changes after ten years, and according to the possible changes after ten years, make the integration of production and education, and introduce the talents of head enterprises into schools to teach (Wu, 2022).

2.2 NEE Schemes at Local and Institutional Levels

Following the implementation of the new engineering education (NEE) strategy at the national level in China, numerous local universities have responded by launching corresponding NEE reform schemes. Renowned institutes of science and engineering, such as Tianjin University, University of Electronic Science and Technology of China, and South China University of Technology, have initiated their own new engineering construction schemes. Moreover, the majority of comprehensive universities have incorporated NEE into their agenda of education and teaching reform.

2.2.1 The Tianjin University NEE Plan 2.0

In 2020, Tianjin University released its NEE Plan 2.0, which aims to deeply reform and comprehensively innovate its engineering education system. The scheme consists of four parts, including guiding thoughts and development objectives, general requirements and basic principles, key tasks and key measures, and organization and implementation. In terms of guiding thoughts and construction goals, Tianjin University seeks to build a new engineering education system with Chinese characteristics that is world-class and characterized by great leadership by 2025. By this time, Tianjin University aims to become a leader in China's higher engineering education and at the forefront of the world's first phalanx of higher engineering education. By 2030, Tianjin University aims to build a new engineering education system with Chinese characteristics that is world-class and characterized by great leadership, become a leader in China's higher engineering education and the world's higher engineering education, and provide the world with Chinese experience in the development of higher education.

Regarding general requirements and basic principles, Tianjin University proposes four general requirements: cultivating people by virtue, highlighting excellence orientation, strengthening engineering innovation, and paying attention to entrepreneurship education. Additionally, Tianjin University puts forward four basic principles, including emancipating the mind and innovating the mechanism mode, taking undergraduate education as the core mission, building a high-level team, and persisting in comprehensively advancing, promoting key breakthroughs, persisting in in-depth exploration, and realizing innovation in teaching and learning.

The most important part of Tianjin University's NEE Plan 2.0 is the third part, which includes four aspects and 12 key measures. The first of the four aspects is to build a complete new engineering education platform system. Specifically, this includes four measures in building an open multidisciplinary training platform, implementing academy system and tutorial system, establishing a scientific teacher appointment mechanism, and deepening multi-disciplinary collaborative education between sectors.

To build an open multidisciplinary training platform, Tianjin University aims to dismantle the wall construction between colleges and majors for the future development of science, technology, and industry. Additionally, Tianjin University seeks to open a new engineering education and training platform at both institutional and departmental levels and explore a new mode of future NEE college. Tianjin University also proposes to dismantle the wall between the university and society, implement deep integration of industry and university, interdisciplinary integration, domestic and international training integration of teaching and research, and active integration. Furthermore, Tianjin University aims to dismantle the wall between teaching and scientific research by realizing the interdisciplinary cooperation among key laboratories, engineering centers, maker spaces, and innovation and entrepreneurship incubators at all levels.

To establish a scientific and practical employment mechanism, Tianjin University seeks to establish an inter-college joint employment system and encourage the establishment of interdisciplinary and inter-college teaching teams on the new engineering school-level platform. This will involve carrying out interdisciplinary and interdisciplinary training programs, formulating curriculum design, especially project-based curriculum design, and promoting the overall improvement of the curriculum level of the whole school by co-constructing and sharing courses.

Lastly, in deepening multi-dimensional collaborative education, Tianjin University requires all kinds of scientific research laboratories at all levels and the engineering center to be fully open. Tianjin University especially emphasizes the active introduction of the whole process of deep participation of enterprises and personnel training, wherein both university and enterprises jointly formulate personnel training objectives, set up a curriculum system, and implement the teaching process of engineering projects in the front line of industry.

The second aspect of the plan is to build a curriculum system with projects as the chain. Tianjin University plans to design, build, and develop three types of projects, namely, the main construction and curriculum project curriculum group project, undergraduate and postgraduate programs, and multi-disciplinary team projects and graduation design. The university also aims to build a science and engineering education curriculum system and a quality education curriculum system that ensures a close cooperation and reasonable connection between courses and projects.

The third aspect of the plan is to carry out flexible and diverse project-based learning. To achieve this aspect, Tianjin University emphasizes dismantling the wall between teaching and learning and encouraging students and schools to carry out problem-oriented teaching modes. The university encourages students to engage in experiential learning, autonomous project practice learning, and infiltration learning.

Through these modes, students can combine the core knowledge of the curriculum with practical applications and develop their ability to solve compound engineering problems, undertake scientific and technology research, and product development.

Finally, the fourth aspect of the plan is to implement the consistent talent training mode at the undergraduate and postgraduate levels. Tianjin University aims to train high-level and specialized talents by implementing the training mode of continuous master's degree or direct reading of master's degree. To achieve this, the university plans to design an integrated talent training program and establish a curriculum system from undergraduate to graduate students. Additionally, the university will construct advanced courses with contents spanning undergraduate.

The fourth major aspect of the proposed plan is to establish a consistent talent training mode for both undergraduate and postgraduate levels. Specifically, Tianjin University intends to adopt a continuous master's degree or direct reading of master's degree training mode to nurture high-level and specialized talents. To achieve this goal, an original integrated talent training program will be designed, along with a curriculum system spanning from undergraduate to graduate levels. Advanced courses with content that spans across these levels will be developed, and credits will be mutually recognized. Additionally, a tutorial system will be established to cultivate students' strong learning ability and outstanding research skills, thereby paving the way for their rapid development.

2.2.2   The University of Electronic Science and Technology of China (UESTC) Plan

The University of Electronic Science and Technology of China (UESTC) is another higher education institution that has implemented the New Engineering Education (NEE) initiative. As an engineering-focused university located in Chengdu, UESTC has developed its NEE program, named "New E3," which is a three-dimensional, six-core, and multi-level system for student cultivation. The program begins with freshmen's project courses, and is based on the knowledge structure of science, technology, engineering, and mathematics (STEM), and fully integrates research and development (R&D) with business to stimulate students' imagination, creativity, and cultivate their business and entrepreneurial skills.

The core of the UESTC NEE plan is to build a project-based course system that starts with freshmen's projects and gradually increases in challenge, while also integrating cross-disciplinary talent training. This plan is perceived as a necessary reform for engineering students that will lead to the future of engineering education at the Chengdu University of Electronic Science and Technology. The program focuses on cultivating curiosity and learning ability, global vision and leadership, and emphasizes both soft and hard skills, including knowledge structure construction, innovative thinking, and value shaping. These efforts aim to enlighten students' thoughts, arouse their curiosity, explore knowledge to stimulate potential, and enhance their personality development.

Since 2016, UESTC has comprehensively carried out challenging research-oriented teaching reforms, focusing on arousing students' curiosity, stimulating their potential, and improving academic challenges. More than 700 high-quality challenging research courses have been built, and the teaching contents, teaching methods, and assessment methods have been reformed. Instructors are encouraged to teach in accordance with research-based pedagogical outcomes, with a focus on changing students' learning methods and guiding them to challenge themselves in learning, ultimately fostering students' innovation competence.

The UESTC NEE plan goes through several important procedures. The first step is to implement the special development plan of freshmen's project-based courses, which is fully incorporated into the undergraduate talent training plan. The university introduces outstanding project-based learning for freshmen to enable them to understand the conception, design, implementation, and operation (CDIO) from the beginning of their studies. By attempting to solve valuable engineering problems, students' curiosity and learning ability are aroused, and they are guided to develop the habit and ability of design innovation, creation, and teamwork. Additionally, through project-based curriculum study, freshmen gain real-world engineering experience, understand the connotation of natural science and professional courses, and learn the relationship between products and courses, enhancing the necessary perceptual knowledge before beginning their professional study. For example, the School of Optoelectronic Science and Engineering of UESTC has set up nine freshman project courses revolving around 3D display and implementation, covering every student in the school.

Secondly, this university has implemented a special plan for designing and practicing the NEE system. Through systematic design and reform, the university has developed a progressive, interdisciplinary, and project-based research course system. The first project, which is a freshman course, introduces students to the major and curriculum system. The talent training system in the second step gradually increases the challenge degree and serves as a hard currency. For instance, the School of Optoelectronic Science and Engineering emphasizes professional interest, cognitive project conception and design ability, project design and implementation ability, project realization and operation ability, and comprehensive integration ability. In addition to the nine freshman project courses, the school offers professional ability improvement project courses based on conceptual design in the first semester of sophomore and junior year, comprehensive design project courses based on design implementation in the second semester of junior year, and multidisciplinary integration challenging courses based on implementation operation in the fourth year. The university ultimately established a step-by-step challenge-oriented and interdisciplinary project curriculum system from first-year project-based courses to design-based project courses, further to challenge-based project courses, and ultimately to project peak courses. In addition, the university achieved full coverage of project-based learning for students across all grades.

Thirdly, the university is implementing a special plan for NEE reform of scientific research-based education delivered by high-level and superior research teams. The core of this plan is to develop and implement a series of challenging project core courses by high-level superior scientific research teams to transfer knowledge from research papers to classroom teaching. For instance, the School of Information and Communication relies on the superior disciplines of information and communication engineering and the two national technological invention awards won by the school to prompt renowned professors to lead a group of young instructors in teaching students with their research outcomes and professional expertise. This plan prioritizes curriculum with projects, constructing a double helix teaching system that connects excellent core courses with excellent customs clearance projects. It also reconstructs the project-based teaching framework and teaching content from junior, intermediate, and advanced to innovation, and elevates the high-level scientific research practice from the second class to the first class with the challenge of case teaching of core courses and customs clearance of scientific research practice. For example, the School of Electronic Science and Engineering relies on the superior disciplines of electronic science and technology, electronic thin films and integrated devices, and the State Key Laboratory, facing key core technologies, creating innovative talents with sensor chips as the core, and cultivating scientific research and education projects. In this project, the sensor chip with multifunctional integration characteristics serves as the source for teaching. Basic science and typical technical problems are extracted from the national key R&D plans and other projects to construct a core curriculum system, implement research projects integrating interest, challenge and display, and develop a step-by-step teaching mode from easiness to difficulty, forming a whole chain research and education framework based on software and circuit chip system demonstration of sensor chip research and development and application demonstration, to realize the integration and interaction between classroom teaching and scientific research practices.

The university has invested significant effort into creating a new ecosystem for cross-border integration of talent training. Specifically, it has developed a series of programs named "smart+" that are focused on future-oriented training. The university proactively identifies new trends in the scientific and technological revolution, industrial transformation, social development, and interdisciplinary integration, and aligns its programs with the national innovation chain, industrial chain, and other strategic development areas. The university has developed a range of highly application-oriented programs, such as artificial intelligence robot engineering, unmanned aerial vehicle system engineering, new energy materials and devices, smart grid information engineering, intelligent manufacturing, Internet financial data science and big data technology, Internet of Things, and digital media technology. These efforts are intended to revamp the talent training curriculum system, enhance the professional certification and evaluation of engineering education, upgrade and transform traditional and advantageous majors, and establish a new engineering program system.

Moreover, the university actively integrates the training system for specific majors with students' soft skills. The university's general education system comprises a general curriculum system and general resource expansion system with six modules consisting of 114 core general courses, introductory courses, and high-level massive open online courses (MOOC). The university has established UESTC Stages, UESTC Halls, and other platforms for students to access typical classical intellectual resources. Every major in the university requires a compulsory course in professional writing and oral expression, which cultivates students' speculative and expressive abilities throughout the four-year teaching process.

The university has also implemented a special training plan for top-notch innovative talents at the university and college levels. The plan emphasizes the integrated "undergraduate-master-Ph.D." degree system, personalized training programs, formation of high-level tutor teams, and establishment of international exchange and learning platforms for students. The university promotes interdisciplinary integration and personnel training mode reform and implements characteristic plans such as Internet Plus Compound Experimental Class, Intelligent Manufacturing Experimental Class, Robot Characteristic Experimental Class, and Electric Vehicle Innovation and Entrepreneurship Compound Experimental Class to implement "Smart + Industry" and "Electronic Information + Industry" categories. The university allows students two opportunities to change majors, without constraints, to cater to individualized learning needs.

The university leverages its resources and advantages to achieve mutual benefit and win–win development, coordinating practical teaching, scientific research, training, science and technology competition, social practice, and other educational activities. The university builds joint research laboratories, cooperative research centers, innovation and entrepreneurship practice platforms, science and technology competition training platforms, off-campus teaching practice bases, extracurricular scientific and technological innovation activity bases, and other educational platforms. It sets up innovation and entrepreneurship colleges, expands collaborative education space, and establishes an innovation and entrepreneurship education system integrating differential teaching, research, and application resources.

2.2.3 The South China University of Technology (SCUT) "F Plan"

South China University of Technology (SCUT) is a science and engineering-focused university that actively responds to the development of the new era of engineering (NEE). In recent years, SCUT has put forward a unique plan named the "F Plan", which focuses on the future industrial development direction and carries out education and training reform of engineering talents in the university. The "F" in the plan represents the future, and it advocates "teaching for the unknown" and "learning for the future", aiming to train every student to become a person who can lead the future in a given aspect.

President Gao Song of South China University of Technology expressed the "NEE F Plan" using a mathematical formula: $F(p) = Pt \times Pl \times Pa$. Among all, Pt stands for the "power of thinking", Pl stands for the "power of learning", and Pa stands for the "power of action". To cultivate independent and critical thinking, SCUT constructs a curriculum system of deep integration of general education and professional education and aims to build up a mature, empowering, and flexible faculty fostering system. To cultivate autonomous learning ability, SCUT stimulates students' curiosity through a teaching system that fosters academic interest. In terms of cultivating action ability, SCUT aims to train students to solve complex problems and explore the existing world or create the unknown world by exploring multidimensional training paths and building a learning community.

To effectively implement the F Plan, SCUT has taken several measures, including the establishment of interdisciplinary colleges. The university's Guangzhou International Campus is being positioned as a pilot area for training leading talents in engineering. This effort is focused on integrating discipline resources around the strategic emerging industries of the country and Guangdong-Hong Kong-Macau Greater Bay Area. Several new engineering colleges have been established with the aim of training students in emerging cross-cutting fields of world science and technology frontier and future industrial development. These colleges are empowered with administrative autonomy to carry out effective enrollment and recruitment mechanisms to attract high-caliber faculty and experts from around the world.

The university also aims to promote cross-integration of disciplines. It seeks to produce innovative and entrepreneurial talents that meet the demands of high-end industries and value chains. Existing engineering majors are strengthened with a focus on science foundations and open sharing. For example, the major of medical imaging trains medical imaging and diagnosis talents through the interdisciplinary study of engineering and medicine.

SCUT is also creating a batch of micro-majors that focus on future technology, future industry, and the digital economy. These micro-majors, such as artificial intelligence and big data technology, offer students the opportunity to take a range of courses and develop their skills in their field of interest.

Plan F recognizes that in modern society, institutions of higher learning are no longer the sole producers of knowledge, and that universities, governments, markets, and the public of enterprises have all become important subjects of knowledge production. To address this, SCUT actively promotes collaborative education among multiple stakeholders and continually improves its talent training mode to meet the demands of NEE.

The implementation of deep integration of science and education is a critical initiative undertaken by this university. To this end, all scientific research platforms have been made available to undergraduates. The university has also strengthened the construction of scientific research infiltration teaching platform, enabling the research of personnel training, and driving high-quality learning. At its Guangzhou International Campus, a new mode of deep integration of science and education has been explored through the establishment of high-end research institutes. The university leverages these institutes to attract high-level international talents, who undertake cutting-edge cross-research and train graduate students. Additionally, the traditional colleges rely on these high-end research institutes to carry out undergraduate teaching and research, and develop new interdisciplinary courses. As a result, a connection is formed that integrates scientific research, teaching, and learning.

Undergraduates participate in professors' research groups in their second year of university, and can take compulsory courses for postgraduates in their junior year. Postgraduates also train basic classroom teaching skills through at least one year's experience as undergraduate teaching assistants, laying a solid foundation for being competent for teaching and research positions in the future.

Collaborative education in this university extends beyond the collaboration between scientific research and teaching. The university has expanded cooperation between schools and enterprises, closely linking with the development of local economy and industry. It introduces enterprise resources and co-builds majors to create a platform for cultivating talents. For example, the university has introduced industrial resources and established microelectronics and software colleges. It has also formed a modular course of university-enterprise cooperation based on engineering cognition. The university has partnered with leading enterprises such as iFlytek, DJI, and others to build joint laboratories and future innovation laboratories. They have set up a scientific conjecture platform with interdisciplinary integration to encourage students to explore the unknown. The university also collaborates with leading enterprises to set up characteristic teaching reform classes. The schools and enterprises jointly formulate training objectives, build a curriculum system and teaching content, implement training processes to evaluate training quality, and jointly train new engineering education talents (Gao, 2019).

The university places a high priority on promoting the integration of science and education and production and education. In line with this, special attention is given to cultivating the professional and technical ability of its instructors. The university strongly supports and requires every instructor to guide at least one innovation and entrepreneurship project and accept at least one undergraduate student in each employment period. Teachers are encouraged to build their classrooms on enterprises and production lines to improve students' practical skills and innovation ability. To further enhance the university's efforts, entrepreneurs, industrial investment experts, and senior engineers are successfully invited to the podium to offer their perspectives. The university has also set up university-enterprise joint innovation classes and collaboratively developed courses, aimed at fostering closer ties between academia and industry. Through these initiatives, the university seeks to prepare students for the rapidly evolving demands of the job market, while also strengthening its research capabilities and industry partnerships.

2.3   Major Areas of China's Engineering Education Reform

Whether at the national level with new engineering education development policies, or institutional level with NEE advancement schemes, there are commonalities in terms of areas of engineering education reform across discourses and policy texts.

Firstly, many higher education institutions, guided by Ministry of Education policies, consider university-industry collaboration as a crucial means of improving higher education quality. Against the backdrop of Industry 4.0, industry partners are seen as important agents in transforming teaching and learning patterns on campus, and in turn, improving student outcomes. The Ministry of Education and other governmental bodies have issued policies to coordinate the relationship between the industry and higher education sectors, and have implemented incentive mechanisms to reward enterprises that participate in university development. Different higher education institutions have also established specific mechanisms to welcome industrial partners to enhance higher education teaching and learning quality. University-industry partnership has been regarded as a source of innovation for teaching and learning.

Another important aspect of China's engineering education reform is the establishment of Modern Industrial Colleges at universities. The projects of industry-university cooperation and collaborative education conducted by the Ministry of Education and the newly developed Modern Industrial Colleges differ in depth and breadth, as the latter involves complete and varied industrial elements within higher education systems. In the modern industrial colleges, enterprise representatives reside at universities for long-term periods, and industrial cooperation extends beyond micro-level curriculum and teaching to overall professional scheme design, equipment procurement, and practice training.

In addition to seeking external help from the industry sector, policy texts across ministries and institutions emphasize the responsibility of higher education institutions to initiate teaching and learning reforms within their campus settings. Active learning, experiential learning, evidence-based learning, and other approaches consistent with pedagogical research outcomes have been identified as necessary for students to take initiative in their learning. Thus, many institutions prioritize project-based learning in the process of reforming teaching and learning. Teaching instructors with reference to context, industry, and engineering professionalism has been deemed an important mission to make faculty members more proficient in teaching students authentic and real-world engineering problem-solving.

Structurally, building new majors and transforming existing majors are important ways for many colleges and universities to construct new engineering courses, in line with the needs of industrial development under the new situation. The past few years have seen the revocation and reconstruction of many programs in Chinese universities.

The following chapters will address these major engineering education reform areas in detail.

References

Gao, S. (2019). Implementing F plan to cultivate quality engineering talents. *Higher Engineering Education Research, 4*, 19–25.

Koh, A., & Zhuang, T. (2021). "New engineering education" in China: A national technological imaginary of the Chinese Dream. *Asia Pacific Education Review, 22*(1), 31–40.

MOE. (2017a). *Fudan consensus for new engineering education development (in Chinese)*. https://baike.baidu.com/item/%E2%80%9C%E6%96%B0%E5%B7%A5%E7%A7%91%E2%80%9D%E5%BB%BA%E8%AE%BE%E5%A4%8D%E6%97%A6%E5%85%B1%E8%AF%86/20612927?fr=aladdin

MOE. (2017b). *Tianda action for new engineering education development (in Chinese)*. https://jys.tlu.edu.cn/_upload/article/files/5d/e4/7a8bdabe4381a8dc18b0cbf45af5/020bc7e8-fd44-44a5-8a7d-b00333266f01.pdf

MOE. (2017c). *Beijing compass in new engineering education*. http://www.moe.edu.cn/jyb_xwfb/gzdt_gzdt/moe_1485/201706/t2017c0610_306699.html

Wu, Y. (2022). *Deepening the development of the 'Four New Project' to pursue a path of talent self-cultivation*. http://heri.cau.edu.cn/art/2022/6/28/art_37510_868897.html

Zhuang, T., & Xu, X. (2018). 'New engineering education' in Chinese higher education: Prospects and challenges. *Tuning Journal for Higher Education, 6*(1), 69–109.

Chapter 3
Advancing a Synergistic Approach to Engineering Education Through the Massive Teaching-Focused University-Industry Collaboration

Abstract This chapter provides a detailed examination of the manner in which the government, under the auspices of the emergent macro educational policy for engineering disciplines, assumes a pivotal role in orchestrating the synergies between the industrial sector and higher education institutions. The governmental intervention has been instrumental in catalyzing collaborations between academic institutions and corporate entities, with a specific emphasis on educational objectives. Available empirical evidence suggests that the magnitude and frequency of industry-academia partnerships in China have undergone marked augmentation in recent years, attributable to the concerted endeavors of governmental bodies, academic institutions, and the broader industrial community. Furthermore, this chapter offers an in-depth analysis of the salient models through which the industrial realm has facilitated the enhancement of higher education quality and probes the multifaceted strategies implemented in manifesting the first mission of higher education.

Keywords Synergistic approach · University-industry collaboration · Teaching-focused collaboration · First mission of higher education · Educational innovation ecosystem

Despite the lengthy history of university-industry collaboration (UIC) in the global higher education field, many of these collaborations focus on non-educational issues such as collaborative knowledge transfer, patent recognition, and collaborative research, to name a few. Conversely, the number of industry-university collaborations geared towards education and teaching is quite limited.

In recent years, China has experienced rapid development of UIC for teaching purposes, aiming to be a crucial means for enhancing the quality of engineering education. This collaboration between the two sectors is referred to as teaching-focused UIC or education-focused UIC (Borah et al., 2021; Zhuang & Shi, 2022). This type of collaborative educational project between industry and universities aims to deepen the integration of industry and education, promote the organic connection of education, talent, industrial chain innovation, and talent training

T. Zhuang, *Modernizing China's Undergraduate Engineering Education Through Systemic Reforms*, SpringerBriefs in Education,
https://doi.org/10.1007/978-981-99-6388-1_3

reform in universities, aligned with the latest needs of industrial and technological development.

For universities, cooperation with industry for teaching purposes can break the bottleneck of personnel training, making personnel training closer to industrial needs and social reality, and transform more social resources into educational resources, thereby optimizing environmental conditions for practical education. Enterprises can transfer the construction of their talent team to talent training, collaborate with universities to cultivate outstanding talents urgently needed by the industry, gain a first-mover advantage in enterprise talent ecological construction, and industrial upgrading and innovation, providing robust support for sustainable enterprise development.

UIC for teaching can be traced back to 2014 at the earliest in China, but this cooperation mode has flourished since 2017. According to an internal journal on industry-university cooperation run by the Ministry of Education, in 2014, only eight enterprises in China started carrying out industry-university cooperation for teaching purposes, including Baidu, Google, Intel, IBM, Microsoft, Siemens, Beijing Touch Technology, and Original Computing. In 2015, only 15 enterprises cooperated, and a total of 243 similar industry-university cooperation projects were set up with universities to provide financial support, amounting to a total of 17 million RMB. Although the number of participating enterprises and the overall number of projects increased in 2016, it still belonged to a relatively slow development stage. Since April 2017, this mode of industry-university cooperation has witnessed a spurt of growth.

In 2017, the participation of enterprises and universities increased significantly, with a total of 218 enterprises cooperating with 892 universities to establish nearly 10,000 software and hardware support provided by enterprises, with about 2.316 billion RMB and financial support of about 552 million RMB. In 2018, the Ministry of Education launched a platform for collaborative education projects between industry and universities to realize standardized management of the whole process of such projects. On this platform, a total of 1090 undergraduate universities cooperated with 489 enterprises to establish 17,654 projects throughout the year, with enterprises providing software and hardware support amounting to about 7.143 billion yuan, and financial support of about 641 million yuan.

The types of universities participating in this kind of industry-university cooperation now cover the deployment and joint construction of universities, local universities, military academies, and overseas universities. Many well-known enterprises at home and abroad, such as Huawei, Tencent, Baidu, Inspur, and Google, actively participate in industry-university cooperation for teaching purposes, which has led to the production of a multitude of new teaching materials, courses, and systems that reflect the latest progress in industry and technology, transforming more high-quality social resources into educational resources. The corresponding quantitative indicators have further increased in the following three years.

In 2020, a total of 524 enterprises participated in the declaration of intellectual property rights cooperation projects, of which 333 successfully passed the assessment. In the second batch, 556 enterprises participated in the declaration, with 366 passing the assessment. Ultimately, 16,717 projects were approved, and the enterprises provided approximately 2.952 billion yuan in software and hardware support and 559 million yuan in financial support. The number of scientific and technological enterprises and industrial enterprises participating in the projects is on the rise. Notably, many universities affiliated with central ministries and commissions also participated in the program, and over 60% of the first-class universities were involved (Shi et al., 2022).

3.1 Government Coordination Between the Industry Sector and the Higher Education Sector

Industry and higher education are two sectors with distinct attributes. The primary objective of the industry is to achieve short-term economic benefits, which demands visible and tangible returns in a relatively brief period. However, the purpose of higher education is mainly to cultivate human potential, and therefore, it is unlikely to achieve conspicuous outcomes in a short period. Consequently, the collaboration between enterprises and higher education institutions cannot occur automatically and requires mediation by third parties or other stakeholders.

In China, the collaboration between industry and universities for educational purposes began with the relevant notice from the Ministry of Education. With the continuous improvement of the corresponding mechanism, the development of industry-university cooperation projects for the purpose of teaching has formed a standardized process with the year as the basic unit. The concrete matching and implementation follow several steps described below.

Since 2017, the Ministry of Education has released an annual public notice to solicit enterprise guides. The notice specifies and explains the essential conditions for enterprises to participate, the types of industry-university cooperation projects supported by key points, the direction of key areas supported by key points, and some specific work processes. For instance, the Ministry of Education stipulates that participating enterprises in industry-university cooperation with collaborative education projects must have independent legal personality, advanced technical strength, and R&D capabilities in their industries and fields. Additionally, these enterprises should have stable business operations, a strong financial system, and a good credit record without any instances of deception or fraud. Moreover, their registered paid-in funds should be more than RMB 5 million in principle. In related fields, the enterprise should have a solid foundation for cooperation with universities, and it is recommended to have two or more experts with senior titles in cooperative universities issue recommendation materials.

The Ministry of Education encourages enterprises to provide at least 500,000 RMB in actual support funds to universities in each batch, excluding investments in software and hardware, among others. The actual support funds provided by enterprises will be used as special funds for collaborative education projects of industry-university cooperation without additional conditions. The Ministry of Education clearly stipulates that the cooperative education programs between universities and enterprises should include six categories (Zhuang & Zhou, 2022).

The Ministry of Education in China has established a standardized process for industry-university cooperation projects for teaching purposes, with the year as the basic unit. Enterprises and higher education institutions have different goals, with the former typically focused on pursuing short-term economic benefits and tangible returns, while the latter aims to cultivate individuals. Therefore, collaboration between these two sectors often requires the mediation of third parties or other stakeholders.

To initiate industry-university cooperation, the Ministry of Education issues a public notice annually inviting enterprise guides to participate. The notice specifies the basic conditions for enterprises to participate, the types of industry-university cooperation projects supported, the direction of key areas, and specific work processes. For instance, participating enterprises must have independent legal personality, advanced technical and R&D strength in their respective industries and fields, and stable business performance. Additionally, the enterprise should have a sound financial system and good credit without any bad records of deception and fraud. Participating enterprises should provide no less than RMB 500,000 in actual support funds to universities in each batch, which will be used as special funds for collaborative education projects without additional conditions.

The Ministry of Education has categorized the industry-university cooperation projects for teaching purposes into six categories. The first category involves projects on New Engineering Education, New Medical Education, New Agriculture Education, and New Humanities Education. Enterprises must provide funds and resources to support research and practice in these areas and promote school-enterprise cooperation. The second category is on the reform of teaching content and curriculum system. Enterprises need to provide funds, teachers, technology platforms, and the latest industry requirements for personnel training to promote updating of teaching content and improving the curriculum system. The third category involves teacher training programs to enhance their teaching level and practical ability.

The fourth category pertains to industry-university cooperation projects for practice conditions and practice base construction. Enterprises are required to provide funds and software or hardware equipment to support the construction of laboratories, practice bases, and practical teaching resources in colleges and universities. The fifth category involves innovation and entrepreneurship education reform projects, which require enterprises to provide software and hardware conditions and investment funds to support the construction of innovation and entrepreneurship education curriculum system, practical training system, maker space project incubation, and transformation platform. The sixth and final category is the joint fund project for innovation and entrepreneurship, which guides teachers and research directions to support college students in innovation and entrepreneurship practice. In each specific project, the support funds provided by enterprises should not be less than RMB 20,000.

The Ministry of Education has stipulated several practical directions for industry-university cooperation, which are supported annually. In 2021, the key support directions include industry-university cooperation and collaborative education projects in the fields of artificial intelligence, digital economy, robots and equipment manufacturing, new energy, intelligent agriculture, intelligent medicine, and new liberal arts.

Regarding artificial intelligence, the Ministry of Education expressed support for both schools and enterprises to conduct research on deepening the connotation of artificial intelligence and building a training system that emphasizes basic theory, talents, and artificial intelligence plus X compound talents. Through industry-university cooperation and collaborative education, the Ministry of Education aims to explore the teaching content, curriculum construction, teacher training, and innovation and entrepreneurship education reform of artificial intelligence majors and related majors that are closely integrated with industrial development.

The Ministry of Education supports active research on industry-university cooperation in the field of intelligent manufacturing of intelligent vehicles by robots, with the aim of training compound talents proficient in the basic theory and professional knowledge of robots, as well as cross-integrated engineering and technical talents engaged in intelligent manufacturing design services. This will improve the personnel training level in various fields of artificial intelligence.

In the field of digital economy and information industry, the Ministry of Education focuses on supporting the training of high-level information technology talents who can quickly adapt to the requirements of digital industrialization. Key support areas include big data, cloud computing, Internet of Things, cyberspace security, blockchain, and 5G communication integrated circuit basic core software. To serve new engineering majors in information technology fields such as science and engineering, both schools and enterprises are required to closely carry out curriculum and teaching resources construction, organize relevant teacher training, innovate personnel training mode, and jointly build general courses in information technology fields between schools and enterprises.

In the field of robot and equipment manufacturing, the Ministry of Education focuses on supporting both industry and university to promote the deep integration of new generation information technology and manufacturing technology around the strategy of manufacturing power. This will accelerate the training of compound applied and innovative talents in the field of robots and intelligent equipment. The sub-directions mainly supported include industrial robots, service robots, intelligent robots, and aviation equipment, development of courses and teaching resources in the fields of rail transit equipment, offshore engineering equipment, satellite and its application industry, intelligent manufacturing equipment, etc. Additionally, the Ministry encourages the construction of relevant experimental environment through virtual simulation.

Regarding new energy, the Ministry of Education focuses on supporting the development of energy storage technology. Enterprises are encouraged to actively participate in the development of teaching materials for professional planning personnel training programs in corresponding schools. Efforts in teaching design, curriculum setting, practice training, etc. will promote the needs of enterprises, integrate into personnel training, set up core curriculum groups of related majors, and explore tasks facing the real production environment of enterprises as training modes. The Ministry also emphasizes the importance of strengthening the construction of students' practice bases and teachers integrating production and education.

With respect to intelligent agriculture, the Ministry of Education is focusing on supporting the construction of a practical teaching platform for the intelligent agriculture specialty based on stations. In view of the gap between the training of talents in agricultural universities and the demand for talents in the agricultural industry, the Ministry aims to create a number of service industries built in production areas that adhere to domestic first-class practical education resources with various functions for an extended period. The goal is to form a high-level practical teaching team of science and technology promotion with multi-disciplinary participation and a combination of full-time and effective resources. The Ministry aims to train professionals who can master and apply modern information technologies such as the Internet of Things, big data, and artificial intelligence to engage in research and development and application of smart agricultural machinery and smart agricultural systems.

In terms of intelligent medicine, the Ministry of Education is encouraging enterprises to support the construction of related specialties in the field of intelligent medicine. The goal is to train professionals who can work with intelligent medical instruments and surgical robots engaged in intelligent medical treatment and telemedicine, intelligent medical imaging and intelligent diagnosis, patient-centered intelligent health management systems, intelligent drug mining and medical research, virology, and etiology research.

From the above paragraphs, it is evident that the Ministry of Education first defines the key areas and general directions for the entire cooperation before matching industry and universities. Without such coordination and active calls, it would be challenging for universities and enterprises to spontaneously seek each other out and identify each other's strengths to establish a suitable partnership. It is also unlikely that enterprises will make significant investments in higher education, which is not closely related to profit-seeking. The issuance and call of the Ministry of Education is merely the first step of industry-university cooperation, and the actual matching and cooperation implementation require specific steps before they can be guaranteed.

3.2 Selecting Appropriate Partners

Next, the government holds an annual or biannual on-site matchmaking conference for industry-university cooperation, providing a platform for numerous business circles and universities to understand each other's advantages and strengths before selecting suitable partners. At these matchmaking meetings, government representatives, enterprises, and universities can communicate face-to-face and openly ask questions. Enterprises will be invited to demonstrate their strengths and advantages, such as the reform of education modes or technical products, on such a platform.

Enterprise leaders will explain in detail how their technology and products can effectively improve the quality of higher education. Each university can evaluate whether an enterprise has the necessary ability and advantages to help improve the quality of its education and teaching reform. University representatives are also invited to share their demands for strengthening certain aspects of teaching and learning and the resources they need to achieve their goals. Sufficient free time is provided for representatives from both sides to get to know each other.

Enterprises interested in participating in these projects must first submit their business guide to the Ministry of Education, which articulates their technical, technological, and economic strengths in areas where they can provide assistance to improve university education. In the project guide, enterprises must specify which of the six categories of projects they want to establish as required by the Ministry of Education. They must also clarify a wide range of other issues, such as their advantages, the connection between enterprise technology and content, and the willingness to invest in industry-university cooperation projects for the long term. Enterprises will also put forward corresponding requirements for partner colleges and universities, including what proportion of colleges and universities should embed the technology and content provided by enterprises into their own curriculum system and what kind of certificates colleges and universities should issue at the end of cooperation projects to demonstrate that they used the help of enterprises in the process of curriculum reform or teaching reform.

After the enterprise submits the guide, the Ministry of Education organizes an expert group to review the various guides in detail. The Ministry of Education will review whether the applicant meets the minimum requirements and whether the applying enterprises are qualified to participate in such collaborative projects. The expert group rejects corresponding applicants who do not meet the qualifications. For qualified enterprises, the Ministry of Education will disclose their list in batches at specific timepoints every year to the general public.

After seeing the list of provided enterprises and being asked by the Ministry of Education, higher education institutions will choose potential partners on a specialized platform to enter the initial collaboration stage. The most critical selection criteria for colleges and universities are their specific needs to reform their teaching, learning, and personnel training. In this process, representatives of universities and enterprises often go through multiple rounds of communication, confirmation, and reconfirmation before entering the corresponding contract signing stage.

After signing the agreement, universities and enterprises commence formal cooperation, which often lasts for two years. At this time, the main role of the government is one of supervision and support. The specific forms of cooperation will vary depending on the types of industry-university cooperation projects. Some projects focus on teacher training, while others focus on the reform of teaching modes and methods. Some projects provide students with real industrial experience, and others focus on improving existing methods of experimental teaching. According to the contents of the contract signed with enterprises, higher education institutions incorporate the product technology or schemes from the industry sector into their course system, while enterprises provide the agreed-upon technology or any other tangible outcomes. Further communication may be necessary if unexpected outcomes occur concerning the degree of students' acceptance of elements from the industry sector.

3.3 Striving for a Higher Education Innovation Ecosystem

Enterprises participating in industry-university cooperation generally collaborate with various universities, with some partnerships covering multiple levels and types. For instance, renowned companies such as Huawei, Alibaba, IBM, and Texas Instruments have partnered with dozens of higher education institutions in China, covering several or all of the six categories of projects stipulated by the Ministry of Education. In turn, different majors and fields in various colleges and universities choose different enterprises as partners. Therefore, under the promotion of the Ministry of Education, the industry-university cooperation between enterprises and universities has created a cooperative network that is highly interconnected.

In this specific cooperation process, some cooperation projects are solely funded by enterprises, while others are jointly funded by both parties. Some projects are relatively elementary, while others are more profound. In fact, this overall pattern of full participation, multi-faceted participation, and multi-level participation is the primary goal of the Ministry of Education in carrying out industry-university cooperation projects.

Teaching-focused university-industry collaboration projects differ significantly from other national projects in China, such as the "Double World-class University" campaign, which often involves only the top higher education institutions in China and not all higher education institutions. In the "Double World-class University" campaign, institutions such as Tsinghua University, Peking University, Shanghai Jiaotong University, Nanjing University, and Zhejiang University are often in the limelight. However, it is essential to note that these top institutions are only a small part of China's higher education landscape, representing the top quality but not the average level of China's higher education sector.

In recent years, the collaborative education project of industry-university cooperation has involved colleges and universities of all tiers. Participating institutions include not only the most first-class key institutions but also a large number of second-rate and third-rate ordinary undergraduate institutions and even colleges. This indicates that institutions at all levels, not only the best ones, can receive support from enterprises to promote their teaching and curriculum changes. With such a large-scale guarantee, the influence of industry on higher education can be said to be comprehensive and holistic.

While universities and enterprises are the primary actors in the process of industry-university cooperation, the government has also promoted several practical policy levers to foster an innovative ecosystem (Zhuang & Liu, 2022). For instance, the document of the General Office of the State Council on Several Opinions on Deepening the Integration of Production and Education suggests that the state encourages financial institutions to support university-industry collaborative education projects in accordance with the principles of controllable risks and sustainable business. Moreover, the state supports the construction of qualified production-education integration projects by using China's government-enterprise cooperative investment fund, loans from international financial organizations and foreign governments, and even the Silk Road Fund of the Asian Infrastructure Investment Bank. The state guides the banking industry and financial institutions, innovates the service mode, and develops diversified financing varieties suitable for the characteristics of the integration of production and education projects, providing financial support for the cooperation mode between the government and social capital. The state also encourages qualified enterprises to issue standardized creditor's rights products through equity financing in the capital market, increases investment in production-education integration training base projects, accelerates the development of student internship liability insurance and personal accident insurance, and encourages insurance companies to set special rates for new apprenticeship insurance for modern apprenticeship enterprises.

3.4 Means Through Which the Industry Sector Helps Boost Higher Education Quality

In this section, we will examine selected cases of industry-university cooperation chosen by the Ministry of Education to illustrate how enterprises can contribute to improving the quality of higher education in the specific context of such cooperation. In September 2021, the Ministry of Education organized an annual matchmaking conference on collaborative education between industry and universities in Beijing, where it announced 124 exemplary cases of industry-university cooperation. Of these, 57 cases were specifically targeted at reforming teaching and curriculum content. The Ministry of Education's excellent cases of industry-university cooperation not only list the cooperative institutions, enterprises, and projects, but also provide detailed descriptions of the main achievements, project promotion, and applications for each case through a dedicated website, which is supported by written introductions and photographs of real-life examples.

3.5 Jointly Forming a Comprehensive Teaching Reform Plan with Colleges and Universities

In this section, we examine exemplary cases of industry-university cooperation selected by the Ministry of Education to demonstrate how enterprises can contribute to improving the quality of education and teaching through this specific collaborative process. In September 2021, the Ministry of Education hosted an annual matchmaking conference on collaborative education between industry and university in Beijing, where it announced 124 typical excellent cases of industry-university cooperation. Among them, 57 cases were directly aimed at reforming teaching and curriculum content. These excellent cases of industry-university cooperation not only include a list of cooperative institutions, enterprises, and projects but also provide detailed introductions and real-life pictures of the main achievements, project promotion, and application for each case on a specially built website.

The comprehensive teaching reform plan developed by enterprises and universities is primarily based on two aspects of assisting universities in updating the teaching mode and optimizing the teaching system. Regarding updating the teaching mode, enterprises carry out two main tasks. Firstly, they help universities innovate teaching modes based on actual production needs. Secondly, enterprises use advanced technological means to act as resource and technology providers and assist universities in establishing mixed teaching models. In the process of innovating teaching models based on actual production needs, enterprises play the role of "subject" by guiding colleges and universities to modify existing teaching models to enhance the connection between college teaching and enterprise production.

For instance, in the excellent cases selected by the Ministry of Education, Hengyang Normal University integrates the technical advantages of cooperative enterprises in the fields of chemistry and chemical engineering into the chemical engineering professional fields of the university, introduces and embeds the core courses of the industrial chain, and implements case teaching. Huawei Technologies Co., Ltd. focuses on solving practical problems and helps Tianjin University innovate its teaching organization form to form a learner-centered, problem-based chain training model to address the disconnection between curriculum and practice content and the latest technology in the industry. Guanghui City (Chongqing) Science and Technology Co., Ltd. assists Shandong University of Architecture in establishing an "online" graduation design teaching organization system, and supports students majoring in urban and rural planning to complete online graduation defenses. Shenzhen Dingyang Technology Co., Ltd. helps Xi'an Jiaotong University develop a mixed teaching mode of "teacher explanation + experimental micro-class", promoting the teaching reform of the course.

In terms of optimizing the teaching system, enterprises primarily help colleges and universities improve existing teaching systems by providing assistance to make them more scientific and reasonable. For example, Beijing Runnier Network Technology Co., Ltd. helped Sichuan Normal University update its teaching methods and evaluation system based on the experimental teaching platform, while Baidu Online Network Technology (Beijing) Co., Ltd. helped Beijing University of Aeronautics and Astronautics build a teaching system of professional compulsory courses and reconstruct the teaching content.

3.6 Preparing and Updating Teaching Materials

Assisting colleges and universities in carrying out teaching content reform is another avenue through which enterprises can offer support. One way that enterprises can provide assistance is by preparing or equipping teaching materials. During this process, enterprises primarily aid colleges and universities in the preparation of teaching materials for school-enterprise cooperation projects, or for specific subjects and courses required by the school. For instance, Alibaba Cloud Computing Co., Ltd. and Southeast University released supporting teaching materials based on cooperation projects, while Harbin Institute of Technology and Beijing Green Building Software Co., Ltd. established course guidance teaching materials for "Green Building Acoustics" and also have plans to publish new course materials.

Enterprises also help optimize the practicality and professionalism of teaching materials by providing teaching cases and various materials with distinctive industry characteristics. For instance, Nanjing Yanxu Electric Technology Co., Ltd. and Henan University of Technology jointly developed typical teaching cases on new energy vehicle technology, while Shandong Inspur Zhuyuan Education Technology Co., Ltd. provided big data teaching tools for Tianjin University, relying on the company platform. In another example, Wuhan Guangchi Education Technology Co., Ltd. and Chongqing University of Posts and Telecommunications collaborated to prepare supporting teaching materials on optoelectronic information science and engineering.

3.7 Carrying Out University-Enterprise Joint Teaching

In order to bring college teaching closer to the practical demands of the industry and engineering situations, and to enhance students' comprehension of industry practices, the implementation of school-enterprise joint teaching stands as a crucial approach for enterprises to promote reform in the college teaching content system. School-enterprise joint teaching can be classified into two main forms: the integration of enterprise production content into university classrooms, and the collaborative teaching between university teachers and enterprise experts. Regarding the integration of enterprise production content into university classrooms, cooperative enterprises assist colleges and universities in reflecting on the original teaching content, actively incorporating advanced theoretical knowledge and practical experience from cooperative enterprises, supplementing the necessary teaching content required by cooperative enterprises, and ultimately improving the external relevance of the teaching content. For instance, Nanjing Normal University has allowed Baidu Online Network Technology (Beijing) Co., Ltd. to participate more fully in the pre-class, in-class, and post-class teaching processes, introducing the latest technology and excellent resources from cooperative enterprises during project implementation. Beijing University of Posts and Telecommunications has utilized high-quality resources from Huawei Technologies Co., Ltd. in artificial intelligence and the Internet of Things to optimize and embed teaching content into the course material. Concerning the collaborative teaching between university teachers and enterprise experts, universities proactively engage enterprise experts to conduct teaching research alongside their own faculty members. The involvement of enterprise experts in teaching improves the alignment of university courses with actual needs and significantly enhances their cutting-edge and innovative nature. For example, Dalian University of Technology has employed experts from Huawei Technologies Co., Ltd. to participate in teaching and integrate enterprise cases into classroom instruction. Harbin University of Technology has employed experts from Beijing Green Building Software Co., Ltd. and other enterprises and research institutes to co-organize courses.

3.8 Building and Innovating the Teaching Platforms

The use of high-quality teaching platforms can provide college students with convenient, visual, and situational experimental operations and practices, which can enhance their understanding of abstract theoretical concepts, improve practical abilities, and facilitate deeper comprehension of learning materials. Furthermore, the integration of these platforms into the learning process can enable digital and intelligent evaluation, thereby improving teaching and academic feedback efficiency. Enterprises assist colleges and universities in building and innovating teaching platforms primarily in three ways. First, they help establish the necessary experimental platforms; second, they provide their own technology platforms to achieve teaching functions; and third, they aid in innovating experimental content or practical teaching. In assisting colleges and universities to establish experimental platforms, enterprises mainly establish virtual simulation laboratories or practical platforms tailored to certain courses or their characteristics. These courses are often abstract, theoretical, and complex, making it challenging for students to acquire in-depth understanding through teacher narration alone. However, experimental platforms designed with the support of enterprises enable students to immerse themselves in the scenario and visualize and concretize the abstract theory through self-operation, thereby facilitating better comprehension and knowledge mastery. For instance, Beijing Bodao Qiancheng Information Technology Co., Ltd. helped Fuzhou University establish a practical operation platform for the course "Business Data Analysis and Application" to provide comprehensive training in data analysis for students. Chengdu Taimeng Software Co., Ltd. and Tianjin University jointly established a virtual simulation experiment project to assist students in exploring analog circuits, physiology, and other relevant courses in-depth. Regarding the direct provision of teaching platforms to colleges and universities, enterprises function as "suppliers," providing efficient teaching platforms that enable colleges and universities to update their teaching models and provide students with ample practical operation opportunities. For example, Hebei University of Technology leverages the teaching platform of Beijing Shidai Xingyun Technology Co., Ltd. to establish a smart experiment pocket laboratory that allows students to independently design circuit experiment programs. Jiangxi University of Finance and Economics employs the teaching platform of Beijing Shiji Superstar Information Technology Development Co., Ltd. to revamp the teaching mode of insurance courses, enhance student engagement, and broaden their perspectives. Finally, enterprises also collaborate with colleges and universities to develop innovative experimental projects in teaching platforms. For instance, Shenzhen Dingyang Science and Technology Co., Ltd. helped Xi'an Jiaotong University design comprehensive experimental and research projects on electrotechnics.

3.9 Developing and Designing New Courses

As a principal body with advanced technology, resources, and cutting-edge technological dynamics, enterprises play a significant role in promoting the design and development of professional courses in colleges and universities in various forms, such as online, offline, or hybrid courses, through assistance and cooperation. Concerning assisting curriculum development, enterprises mainly adapt to and promote the design and development of new curriculum elements in the existing curriculum system of colleges and universities. As resource providers and technology supporters, enterprises offer strong support for the independent development of design courses in colleges and universities.

For instance, Dalian Tongke Applied Technology Co., Ltd. has aided in the construction of Muke in the Practice of Electronic Instruments at Dalian University of Technology, while Beijing Xiangxinli Technology Co., Ltd. has supported the development of the "Golden Course" in the Risk Assessment of Transmission Lines under Wind Disaster at Wuhan University of Technology. These universities have plans and preliminary measures to carry out the construction of relevant courses. After entering the collaborative state, the two enterprises have input their technical elements, effectively improving the quality of relevant courses. Moreover, Changzhou University utilized the resources of Huike Education and Technology Group Co., Ltd. to construct provincial first-class courses and Muke, while Hubei University of Technology employed the BIM technology and series software platform of Guanglianda Technology Co., Ltd. to develop the training course of Construction Organization Design. The cooperative development of curriculum primarily involves both schools and enterprises jointly promoting the creation and construction of new curriculum. For instance, Hubei Academy of Fine Arts cooperated with Brilliant City (Chongqing) Technology Co., Ltd., which provided its MARS platform, to design the course "History of Chinese and Foreign Architecture under the MARS Simulation Architecture Technology" based on the integration of the school's environmental design specialty. China University of Petroleum also collaborated with International Business Machines (China) Co., Ltd. (IBM) to build the training course content combining theory with industrial practical application and cutting-edge technology in the international IT field after a period of communication, run-in, and in-depth understanding.

3.10 Optimizing the Existing Courses in Colleges and Universities

Optimizing the existing courses offered by colleges and universities is another way for enterprises to contribute to teaching and curriculum reform. This optimization involves improving the content and form of courses. In terms of course form, enterprises can promote innovation in teaching methods while retaining the core content of existing courses. This can be achieved by introducing situational, experiential, and interactive elements that facilitate students' cognitive learning processes. Such optimization aligns with the global trend towards student-centered experiential learning. Enterprises can assist colleges and universities in improving the quality of courses, promoting the reform of existing courses, and supporting school-enterprise course integration. Quality improvement involves enhancing the sense of experience and participation of students during learning. For example, Baidu Online Network Technology (Beijing) Co., Ltd. worked with Zhejiang University of Technology and Nankai University to improve the cross-cutting, interactive, and interesting nature of AI courses, thereby enhancing students' ability to solve complex engineering problems and increasing the course's interest and integration. To promote the reform of existing courses, enterprises provide supporting materials for curriculum reform and collaborate with colleges and universities on curriculum framework, objectives, content, and evaluation standards. For instance, Beijing Century Superstar Information Technology Development Co., Ltd. continuously improves Yunnan University's "Construction Law" course by analyzing its online and offline learning data. Baidu Online Network Technology (Beijing) Co., Ltd. and Donghua University collaborated to formulate the objectives and content system of the general AI course. To foster school-enterprise curriculum integration, enterprises introduce cutting-edge industry knowledge and employment standards into colleges and universities' curriculum. Meanwhile, colleges and universities actively integrate cutting-edge elements provided by enterprises to carry out curriculum innovation and improve course quality. For example, Huawei Technologies Co., Ltd. incorporates its enterprise certification training course into Beijing University of Technology's "Internet of Things Comprehensive Training" course, allowing students to apply theoretical knowledge to practical situations. Harbin University of Technology introduces Alibaba Cloud Computing Co., Ltd.'s cloud database to students in undergraduate and graduate courses, enhancing their understanding of the cutting-edge technical features and technical framework in this field.

3.11 Providing Curriculum-Supporting Resources

Enterprises have been playing an important role in the reform of the university curriculum system. By providing continuous support and resources, enterprises can help universities optimize their curricula and make them more relevant to the needs of the industry. This not only benefits the students who receive a more practical education, but also benefits the enterprises themselves by having access to a pool of well-trained talent. One of the main ways that enterprises can assist in the development of the university curriculum system is by developing online learning resources. This can be done in two ways: by developing online course learning websites and by developing course learning platforms. Online course learning websites allow students to access course materials and resources from anywhere, at any time, while course learning platforms provide a more comprehensive learning experience that includes interactive features, assessments, and other tools. In addition to developing online learning resources, enterprises can also provide course supporting resources to universities. This can include courseware, case studies, and training resources that can help students better understand and apply the concepts they learn in class. By providing these resources, enterprises can help universities create a more well-rounded and practical curriculum that prepares students for real-world challenges. Examples of successful partnerships between enterprises and universities include Beijing Normal University's collaboration with Tencent to develop online AI course resources for primary and secondary schools, Shihezi University's partnership with Beijing Century Superstar Information Technology Development Co., Ltd. to build a course learning platform, and Tongji University's collaboration with Alibaba Cloud Computing Co., Ltd. to develop animation courseware on computer system structure. Overall, the assistance of enterprises in the reform of the university curriculum system is crucial for creating a more practical and relevant education system that meets the needs of both students and the industry.

3.12 Providing In-Depth Praxis Opportunities

The significance of practical ability for college graduates, particularly engineering graduates, is indisputable. Studies have shown that experience in engineering projects, a comprehensive understanding of the industry, and the ability to apply learned professional knowledge in real-world scenarios are crucial factors for enterprises when evaluating the proficiency of graduates. Therefore, one of the most critical ways for enterprises to assist colleges and universities in teaching and curriculum reform is by providing in-depth practice opportunities to enhance the practical ability of students. This assistance mainly involves two forms: joint project design and development between enterprises and colleges and universities, and enterprises helping to establish training bases for students.

In the joint project design and development between enterprises and universities, enterprises provide college students with opportunities to work collaboratively with enterprises, enabling students to comprehend the work needs of enterprises, and enhance their problem-solving capabilities. For instance, Nanjing Chaoxiangzhe Network Technology Co., Ltd. and Zhejiang University of Technology conducted a joint design workshop involving multiple disciplines and schools. During this workshop, the cooperation team efficiently produced high-quality design solutions, which allowed the students to improve their practical abilities and deepen their understanding of the work requirements of enterprises.

In helping colleges and universities build training bases, enterprises use their resources to help build specialized training centers based on professional requirements. This assistance plays a critical role in improving students' practical abilities and enhancing the practical teaching proficiency of university teachers. Beijing Weirui Jizhi Technology Co., Ltd., for example, helped Hengyang Normal University build a chemistry and chemical majors' practice base and product R&D center, resulting in a significant improvement in the professional abilities and R&D level of university students. Wuhan University of Technology, Zhejiang Cainiao Chuancheng Network Technology Co., Ltd., and Yum (Catering) Wuhan Co., Ltd. jointly established a talent training base that allows students from the university to practice.

In conclusion, providing in-depth practice opportunities to improve the practical abilities of students is an essential aspect of teaching and curriculum reform, and enterprises can play a critical role in this regard by offering their resources, expertise, and knowledge. Through joint project design and development and the establishment of specialized training centers, enterprises can collaborate with colleges and universities to provide students with the practical skills required to excel in their chosen profession.

3.13 Promoting College Students to Participate in Innovation, Entrepreneurship and Scientific Research Competitions

The content of industry-university cooperation is reflected in the support and guidance provided by enterprises to university students in innovation, entrepreneurship, and professional scientific research competitions. This guidance helps students deepen their understanding of theoretical knowledge and improve their problem-solving abilities. For instance, Baidu Online Network Technology (Beijing) Co., Ltd. supported and selected students participating in relevant competitions in artificial intelligence at the Central South University for Nationalities to foster students' ability to integrate subject knowledge and cultivate their problem-solving skills. Beijing Shixiang Xinli Technology Co., Ltd. guided undergraduates of Wuhan University of Technology to participate in the electric power science and technology innovation competition, which effectively stimulated students' initiative and creativity. Additionally, International Business Machines (China) Co., Ltd. encouraged and supported cloud computing students at South China University of Technology to participate in the research and development of innovation and entrepreneurship projects, and applied for innovation and entrepreneurship projects to promote the reform of the cloud computing talent training system in colleges and universities.

3.14 Diversified Implementers of First Mission of Higher Education

Clark Kerr proposed in his classic book "The Uses of Universities" in the 1960s that the functions of universities mainly include talent training, scientific research, and social services (Kerr, 1964). Kerr's concept of "multiple giant universities" recognizes that the role of universities is to meet the changing economic and social needs. Scholars have compared these three functions to the three missions of higher education. The "first mission" mainly focuses on the human resources function of higher education, which is to cultivate graduates into useful human resources through higher education, and to fulfill the mission of education and talent cultivation (Zhuang & Tao, 2022). The "second mission" emphasizes the important role of higher education in knowledge production, especially in creating knowledge (Pinto et al., 2016). The "third mission" is the result of intensive interaction between universities and governments, industries, and other social stakeholders over the years (Trencher et al., 2014).

The "first mission" of higher education has multi-dimensional objectives, such as humanistic objectives, pragmatic objectives, moral cultivation objectives, and human resources objectives. The social development needs of different periods emphasize different objectives of the "first mission." Early higher education was likened to the "ivory tower" to describe the status of scholars in leisurely and independent research. However, with the rapid development of science and technology, and disruptive innovation, the gap between the actual ability to train graduates in the tower and the actual needs outside the tower has widened. This gap between what is taught, what is learned, and what is needed has become a concern for stakeholders. With the development of the economy and technology, people have higher expectations for the "first mission" of higher education.

The large-scale industry-university cooperation in China in recent years is a direct reflection of the industrial revolution forcing the higher education system to realize reform, innovation, and enhance connotation construction. This round of industry-university cooperation is different from scientific research cooperation or technology incubation cooperation aimed at realizing knowledge transfer. Its goal is not mainly to realize the "second mission" or "third mission" of higher education. Cooperation between enterprises and universities has long existed (Aizpun et al., 2015; Cai & Ahmad, 2021; Camacho & Alexandre, 2019), but most of the cooperation has focused on scientific research or patent cooperation that is more likely to produce tangible results (McCabe et al., 2016; Säisä et al., 2019). The time-consuming and lengthy talent training cooperation is less focused on. This proves that it is more difficult for higher education to fulfill the "first mission" with high quality than the "second mission" or "third mission" because it consumes all the resources that the executive body needs to pay at the micro level. Therefore, the strong transmission of corresponding blessing power from outside the higher education system, such as industrial power, is a key measure to break through the barriers and achieve the overall improvement of the system quality in an open environment.

References

Aizpun, M., Sandino, D., & Merideno, I. (2015). Developing students' aptitudes through University-Industry collaboration. *Ingeniería e Investigación, 35*(3), 121–128. https://doi.org/10.15446/ing. investig.v35n3.48188

Borah, D., Malik, K., & Massini, S. (2021). Teaching-focused university–industry collaborations: Determinants and impact on graduates' employability competencies. *Research Policy, 50*(3), 104172.

Cai, Y., & Ahmad, I. (2021). From an entrepreneurial university to a sustainable entrepreneurial university: Conceptualization and Evidence in the contexts of European university reforms. *Higher Education Policy.* https://doi.org/10.1057/s41307-021-00243-z

Camacho, B., & Alexandre, R. (2019). Design education. University-industry collaboration, a case study. *The Design Journal, 22*(sup1), 1317–1332. https://doi.org/10.1080/14606925.2019.159 4958

Kerr, C. (1964). *The uses of the university.*

McCabe, A., Parker, R., & Cox, S. (2016). The ceiling to coproduction in university–industry research collaboration. *Higher Education Research & Development, 35*(3), 560–574.

Pinto, H., Cruz, A. R., & de Almeida, H. (2016). Academic entrepreneurship and knowledge transfer networks: Translation process and boundary organizations. In *Handbook of research on entrepreneurial success and its impact on regional development* (pp. 315–344). IGI Global.

Säisä, M. E. K., Tiura, K., & Matikainen, R. (2019). Agile project management in university-industry collaboration projects. *International Journal of Information Technology Project Management, 10*(2), 8–15. https://doi.org/10.4018/IJITPM.2019040102

Shi, Y., Gu, S., Li, D., Sun, M., & Li, S. (2022). Data report of MOE educational collaboration projects (2015–2020). *MOE Internal Journal of University-Industry Collaboration, 1*(1), 11–17.

Trencher, G., Yarime, M., McCormick, K. B., Doll, C. N., & Kraines, S. B. (2014). Beyond the third mission: Exploring the emerging university function of co-creation for sustainability. *Science and Public Policy, 41*(2), 151–179.

Zhuang, T., & Liu, B. (2022). Sustaining higher education quality by building an educational innovation ecosystem in China—policies, implementations and effects. *Sustainability, 14*(13), 7568.

Zhuang, T., & Shi, J. (2022). Engagement, determinants and challenges: A multinational systematic review of education-focused university-industry collaborations. *Educational Review,* 1–29.

Zhuang, T., & Tao, Z. (2022). What lessons can university education learn from outside the ivory tower: Insights from engineering alumni. *Teaching in Higher Education,* 1–17.

Zhuang, T., & Zhou, H. (2022). Developing a synergistic approach to engineering education: China's national policies on university–industry educational collaboration. *Asia Pacific Education Review,* 1–21.

Chapter 4
Fueling Changes to Teaching and Learning in University Engineering Education

Abstract The quality of instruction is directly correlated with students' sense of academic achievement and their educational outcomes. This chapter delineates some key reforms implemented at the micro-level of instruction within the context of China's engineering education transformation. These reforms encompass the use of educational technologies like VR and AR to contextualize complex theoretical concepts, the creation of virtual research and teaching labs based on academic disciplines with nationwide reach, and the establishment of large-scale, national-level exemplary courses. Compared to broader macro-policy measures, the content of this chapter is more intricately connected with micro-level reforms pertaining to pedagogy and learning.

Keywords Reform to teaching and learning · Knowledge contextualization and visualization · Virtual communities for teaching-oriented research · Large-scale open and shared online courses

China's reform of engineering education has resulted in tangible changes in curriculum teaching and learning, going beyond mere policy documents. The collaboration between universities and industry, as highlighted in the previous chapter, aims to bring about concrete changes in college students' learning at the micro level. The transformation of teaching and learning is a critical element of the Ministry of Education's policy design, as evidenced by the emphasis on creating "gold courses" and eliminating "water courses" in China's higher education circles in recent years. The Tianda Action, an official document from the Ministry of Education, underscores the need to implement student-centered education, deepen the integration of information technology and teaching, develop and promote online open courses, and leverage virtual simulation and other technologies to innovate engineering practice teaching methods. These policy guidelines have stimulated significant changes in educational practices in recent years. For instance, there has been a massive expansion of virtual simulation technology-based experimental teaching platforms, the establishment of virtual teaching research spaces, and the proliferation of project-based teaching and learning across diverse disciplines.

4.1 Contextualizing and Visualizing Knowledge for Situated Learning

In the last five years, China has made significant efforts to improve the quality of higher education by leveraging information technology. One of the main approaches has been the development of massive virtual reality (VR)-based courses across various disciplines to address the longstanding issues in teaching and learning. Traditionally, Chinese higher education has been criticized for being heavily instructor-centered and lecture-focused, with printed textbooks as the primary learning materials. However, this approach fails to provide a contextualized and visualized representation of complex knowledge points and their interconnections. To address this gap, the Ministry of Education in China has undertaken a massive development of VR-supported interactive and embodied theoretical and experimental courses to enhance students' learning effectiveness. By October 2020, China had accredited over 700 state-level VR courses covering 41 types of academic disciplines, including engineering, medicine, economics, architecture, and education.

The primary objective of developing and implementing VR courses is to make complex content knowledge, particularly theoretical knowledge, visible and easily accessible for students. This approach helps students overcome the abstraction shock that they often experience when learning complex theories. VR technology enables students not only to observe the hidden workings of relevant theories, such as wave transmission but also to engage in immersive experiential learning. For instance, students can set up their own experiment schemes to observe how theories work differently under different technical and parametric circumstances. This interactive and embodied approach to learning increases students' learning agency, allowing them to take a more active role in their education.

VR, as an emerging technology being increasingly applied to the education field (Chen et al., 2020; Lisichenko, 2015), has been shown to effectively provide students with abundant active and experiential learning opportunities that fuel students' creativity and engagement in learning (Bridge et al., 2007). It creates a situation and environment where students are immersed to experience simulated visual, auditory and force sensations revolving around a given field or topic (Ostler, 1994). An increasing amount of sensory experiences and realistic experiences are provided for learners by VR (Wei et al., 2015), and students' understanding of abstract scientific concepts is found to significantly increase with the help of the immersive VR environment (Chen et al., 2020; Yang et al., 2018). Research also shows that learners in a VR-based learning and operation environment have stronger motivation, interest, engagement and critical thinking skills (Parong & Mayer, 2018; Tarng et al., 2017).

The aim is to provide students with the opportunities to experience situated learning, which conceives of learning as a socio-cultural process where both physical and social contexts are critical to a learning activity (Putnam & Borko, 2000). As a strategy for teaching and learning, it holds that authentic and relevant learning happens when learning is grounded within a situated learning context, and effective knowledge acquisition results from the opportunities for learners to learn and live subject matter in the context of real-world challenges (Kelley & Knowles, 2016). As Laver and Wenger (1991) suggest, to "situate" means to locate thinking and doing processes in a particular setting to accomplish certain knowledge and skill tasks. As such, situated learning theory emphasizes the importance of selecting situations that can engage learners in complex, realistic, and problem-centered activities, premised upon that context provides the setting for examining experiences, and learning exists in complex, robust, authentic environments comprising of actions, agency and situations.

According to a state-level special online platform (ilab-x.com) where massive VR-supported courses and simulation experiments are concentrated, there are a total of 3256 such courses developed or co-developed by 2669 higher education institutions, including universities, colleges and post-secondary vocational education institutions. The 3256 courses spread across a total of 61 specialty categories under 11 big fields (i.e. engineering, science, economics, law, education, literature, history, arts, medicine, agronomy, and management) (Table 4.1). Among all, 728 courses have been rated as state-level first-class courses, 1309 selected as provincial first-class course. These courses belong to several different types, such as general course, program basic course, program core course, and other types of courses, which combined cover a total of 26,019 theoretical knowledge points for students to learn in an interactive, experiential and contextual manner.

It is worth noting that those institutions participating in developing and releasing VR-supported courses are not confined to merely premium or first-tier universities in the country, but rather spread across different tiers across geographical locations. Unlike some other national initiatives in which only a small proportion of higher education institutions, mostly first-class universities nationwide, take an active part, the scope of participation in employing VR-supported technology to enhance teaching and course quality in China has been extremely large, encompassing institutions at all levels, known and unknown to the general public. Participating institutions are premium universities (e.g. Peking University, Zhejiang University, Nanjing University), middle-tier universities (e.g. Nanjing University of Posts and Telecommunications, Shanghai University of Finance and Economics, Beijing Jiaotong University) and those much less-ranked institutions (e.g. Jinling Institute of Science and Technology, Wuhan Technology and Business College). They are located in every major region of the Chinese territory, ranging from the most developed east and southeastern region (e.g. Shanghai, Guangdong Province, Jiangsu Province), to middle region (e.g. Henan Province, Hubei Province), and further to the northwestern part of China (e.g. Qinghai Province, Gansu Province), indicating that the development of VR-supported courses and the participation in such a nationwide initiative have been at an unparallel pace and scale.

The following presents a case on the incineration of household garbage and waste in the field of chemical engineering. As an important part of waste handling, effective garbage incineration matters extremely to human well-being and environmental protection, specifically important in preventing urban cities from being encircled by millions of tons of household waste. However, the whole process of waste incineration is systematically complex, involving a series of stages that are not accessible for

Table 4.1 Overall VR-supported course landscape in Chinese higher education

Field	Specialty category	Course number	Knowledge points covered
Engineering (1486 courses, 10,255 knowledge points)	Dynamics	26	288
	Materials	78	640
	Oceanographic engineering	11	107
	Mechanical engineering	175	793
	Armament	20	222
	Aerospace	35	263
	Apparatus	25	226
	Energy and power	38	226
	Traffic and transportation	61	176
	Mining industry	35	354
	Surveying, mapping and geoinformation	15	135
	Civil engineering	124	1092
	Architecture	22	169
	Geological engineering	20	208
	Hydraulic engineering	36	367
	Electric engineering	115	855
	Automation engineering	53	420
	Agriculture engineering	22	163
	Electronics engineering	120	561
	Computer science	50	384
	Forestry engineering	9	76
	Public security technology	20	180
	Textile engineering	14	118
	Bioengineering	64	662
	Environment science and engineering	31	283
	Light industry	23	238
	Chemical and pharmaceutical engineering	121	137
	Food science and engineering	31	240
	Nuclear engineering	22	81
	Biomedicine engineering	28	254
	Security science and engineering	42	337

(continued)

Table 4.1 (continued)

Field	Specialty category	Course number	Knowledge points covered
Science (423 courses, 3165 knowledge points)	Astronomy	9	101
	Chemistry	119	855
	Atmospheric sciences	6	50
	Geophysics	13	128
	Geology	16	148
	Physics	105	798
	Geographical science	26	274
	Marine science	12	134
	Bioscience	105	621
	Psychology	12	56
Economics (91 courses, 854 knowledge points)	Economics	91	854
Law (71 courses, 623 knowledge points)	Law	35	345
	Marxist theory	36	278
Education (62 courses, 667 knowledge points)	Educational science	34	365
	Sporting science	28	302
Literature (93 courses, 887 knowledge points)	Literature	93	887
History (19 courses, 208 knowledge points)	History	19	208
Arts (169 courses, 1623 knowledge points	Arts	169	1623
Medical science (474 courses, 3901 knowledge points)	Basic medical science	123	1159
	Nursing	55	562
	Traditional Chinese medicine	43	399
	Clinical medicine	98	446
	Forensic medicine	12	107
	Pharmacy	53	323
	Medical technology	31	267
	Public health and preventive medicine	59	638
Agronomy (147 courses, 1541 knowledge points)	Zoology	56	690
	Nature conservation and environmental ecology	27	252
	Botany	64	599

(continued)

Table 4.1 (continued)

Field	Specialty category	Course number	Knowledge points covered
Management (221 courses, 2295 knowledge points)	Management science	221	2295
Total		3256	26,019

Source ilab-x.com

ordinary students, such as waste entering the pit and incinerator oven, the furnace incineration, the flue gas denitrification, deacidification and dust removal, leachate treatment, ash toxicology evaluation and stopping the equipment after the incineration. The whole process of the waste incineration is temporally lengthy, difficult to be covered by general teaching contents, and even more difficult for students to operate in person. Almost no universities or colleges directly establish a tangible garbage incarnation plant on or near campus, and very few send their students to a real plant for internships or experiment due to strong health and safety concerns. As such, students have cognitively had vague images of the multiple steps constituting the so-called important overall incarnation process due to the lack of opportunities for on-site learning experiences.

University Z (pseudonym) in Zhejiang Province together with its industrial partners developed a VR-supported waste incarnation system for students enrolled in environment engineering. The VR-based simulation aims to resolve such a learning plight by visualizing critical steps and procedures during the waste handling process, such as incineration, emission purification, percolate disposal, and lime-ash handling. Specifically, for incineration, students are able to observe critical steps including storage pit, separation and sorting, and pit crane soaking. They are allowed to set parameters (e.g. heat value, blast capacity, fire grate) on their own on the VR platform to see differing effects under different circumstances. While students are not permitted to change the property of original waste, which is set as the fundamental data for simulation, students can operate for whether they want to fully sort the waste, partially sort the waste, or extract certain components of the waste. For the stage of emission purification, steps such as heat transfer, selective non-catalytic reduction, desulfuration, deacidification, denitration, and dedusting are allowed to be observed and operated aimed at enabling students to understand the condition on which ultra-low emissions can be realized. By adjusting on the volume of deacidification lime or catalyst, students can observe relevant deacidification and denitration effects. With respect to percolate disposal, the VR platform has an important mission to enable students to understand all the necessary measures for zero emission, encompassing specific steps of sediment, membrane bioreactor, nitration, reverse osmosis, ultrafiltration, and nanofiltration. All the steps are contextualized in the simulated setting of waste disposal plant. Finally, based upon demonstrating the toxicity of the lime-ash, the platform contextualizes and visualizes every necessary step for students to ensure that they understand how dedusting is ultimately disposed in an environmentally-friendly manner.

There are a variety of customized VR facilities developed for students to operate and control, including waste carrying truck, waste storage pit, garbage sorting equipment, grab bucket, incinerator, dust and waste collection system, percolate disposal system, biochemical treatment system, advanced treatment system, selective non-catalyst reductive denitration reactor, rotary spray drying deacidification tower, activated carbon ejector, dry spray calcium, steam-flue gas heat exchanger, flue gas—flue gas heat exchanger, bag filler, wet deacidification tower, cottrell, limestone-gypsum desulfurization tower, chimney, luminescent bacterial toxicity detector, flip type oscillating device, pressure filter, pulverizer, and so forth.

4.2 Development of Discipline-Based Virtual Communities for Teaching-Oriented Research

In the 'smart+' information age, how to improve teachers' teaching ability, how to make teaching resources better, and how to build and share resources for better, remain important questions. Chinese higher education has witnessed the development of discipline-based virtual communities for teaching-oriented research to respond to these questions.

According to Fan Hailin, deputy director of the Department of Higher Education, Ministry of Education, China explores the establishment of virtual teaching-oriented research communities mainly to achieve breakthroughs in three aspects. First, it wants to achieve a breakthrough in the digital development of higher education. Under the impact of the new scientific and technological revolution, Chinese educational circles have realized the learning style of college students, learning content, learning ability and learning form are accelerating reconstruction. The digital development of higher education has become a major strategic issue that affects and even determines the high-quality development of high-cold education, and it is a breakthrough gap and innovative path to realize the quality revolution from learning revolution to high-quality development. The second is to strengthen the breakthrough in the construction of grass-roots teaching organizations. Grass-roots teaching organization is to organize teaching efficiently and carry out teaching and research training. Since the founding of the People's Republic of China, grass-roots teaching organizations in colleges and universities with teaching and research sections as the main body have played an important role in improving teachers' teaching ability in daily teaching management. However, in recent years, the problems of unclear orientation of grass-roots teaching organizations, single form of teaching and research activities, low participation of teachers and imperfect incentive mechanism have become one of the restrictive factors to further improve the quality of personnel training. Carrying out the construction of virtual teaching-oriented research communities is conducive to solving the bottleneck problem in the development of grass-roots teaching organizations in colleges and universities, breaking the time and space constraints, providing a platform for teachers to carry out teaching and research activities with high frequency, high quality and innovation, and re-stimulating the vitality of grass-roots teaching organizations. Thirdly, it hopes to make a breakthrough in the mechanism of continuous improvement of teaching quality. China's higher education sector and China's Ministry of Education have realized that in order to achieve continuous improvement of teaching quality, normal quality monitoring and information feedback are needed. Relying on the virtual teaching-oriented research communities, carrying out activities such as preparing lessons and discussing lessons can enable university teachers to get in-depth and timely feedback from peers in many aspects. At the same time, universities and instructors can also use information technologies such as big data to form a portrait of teachers' teaching development, help teachers understand their own advantages and shortcomings, provide reference for the renewal of teaching concepts, the innovation of teaching methods and the reform of teaching methods, and then realize the continuous improvement of teaching quality (Fan, 2022).

Fan further explained that there are three main tasks in building a virtual teaching-oriented research community in China. The first task is to explore new forms of carrying out teaching-oriented research. Teachers are encouraged to make full use of information technology to carry out interdisciplinary, cross-school and cross-regional teaching and research exchanges, so as to explore diverse and convenient teaching and research modes in colleges and universities and form new ideas, new methods and new forms for the construction and management of grass-roots teaching organizations. The second task is to build a community of teaching development. Relying on the virtual teaching-oriented research communities, teachers are no longer fighting alone, but gathering the collective wisdom of teachers' groups through collaborative lesson preparation, training and exchange, gradually forming a community of teachers' teaching development, and creating a good atmosphere for teachers to return to teaching and love teaching research and teaching. The third task is to build a high-quality teaching resource library. Through various activities of the virtual teaching-oriented research communities, high-quality teaching resources can be continuously precipitated and accumulated, including syllabus, knowledge map, electronic courseware exercises, test teaching, case experiments, training project data sets, etc., so as to promote the co-construction and sharing of resources.

Regarding how to build virtual teaching-oriented research communities, Fan Hailin put forward three types of paths. The first step is to realize classification exploration. At present, China has classified virtual teaching-oriented research communities into three categories: curriculum teaching specialty construction and teaching research reform in terms of content development. In terms of construction scope, virtual teaching-oriented research communities are divided into three categories: campus, regional and national. In terms of disciplines, virtual teaching-oriented research communities cover major disciplines such as science, technology, agriculture, medicine and literature. China hopes that all kinds of teaching and research sections can carry out classified exploration according to the actual situation and

jointly build a new grass-roots teaching organization system with multiple disciplines, types and levels. The second step is to advance step by step. In 2022, China
plans to start the pilot construction of virtual teaching-oriented research communities first. Based on the application feedback of the pilot, the information platform of virtual teaching-oriented research communities are gradually improved.
After the conditions are ripe, it is planned to promote it in a larger scope, launch
a batch of demonstration virtual teaching-oriented research communities in due
course, and strive to realize the networked and systematic construction of virtual
teaching-oriented research communities through 3–5 years' efforts. The third step
is to strengthen synergy. In the construction of virtual teaching-oriented research
communities, teachers' school expert group, technical support department participation, units and other aspects, China hopes that all stakeholders in various departments can effectively link and promote together. Fan, as a representative of the
Ministry of Education, states that instructors are the main body of the construction of
virtual teaching-oriented research communities, so they should give full play to their
initiative, actively participate in the activities of virtual teaching-oriented research
communities, make contributions and gain something. Colleges and universities
are the management units of the construction of virtual teaching-oriented research
communities, so they should earnestly shoulder the work of quality monitoring and
safety prevention and control, and provide supporting conditions for the construction of virtual teaching-oriented research communities. The expert group should
strengthen research and provide guidance and consultation for the construction of
virtual teaching-oriented research communities. Information platform construction
units should constantly optimize technical support, so that the virtual teaching-
oriented research communities information platform can become an online teaching
and research home for teachers, and gradually realize its usability, ease of use and
effectiveness. China's Ministry of Education also hopes that participating units such
as enterprise publishing houses can deeply integrate into the construction of virtual
teaching-oriented research communities, and contribute corresponding strength in
resource co-construction, personnel exchange and achievement promotion.

In fact, the teaching-oriented research communities have had a long history in
China's education system, including both higher education and basic education.
The traditional teaching-oriented research community consists of faculty members
teaching the same subject. They form a pedagogical team and all the team members
work within the same affiliation. Although these teachers can discuss the content of
the subject together and how to use the most efficient methods to teach these subjects,
participants are often limited to specific workplaces and not extended to the broader
academic community. In contrast, the virtual teaching-oriented research community
is responsible for bringing all teachers who teach the same subject together on the
same platform through the media of network technology. For example, even though
a faculty member is working in the least developed western regions in China, he
or she can still participate in the teaching and research activities organized by the
teaching teams of first-class universities in the eastern developed regions through
virtual teaching-oriented research communities, so as to further ensure that their

horizons are at the same level as the high-level teaching quality in the developed regions.

In addition, the establishment of virtual teaching-oriented research communities is also helpful to promote the establishment and development of interdisciplinary teaching and research groups. In the offline physical campus environment, due to the organizational structure of some rules and regulations in higher education such as departments, it is often difficult for many experts and scholars from different disciplines and fields to gather together to discuss how to promote the cultivation and formation of students' interdisciplinary thinking. Such a status quo is what people usually call 'the barriers to disciplinary integration'. The establishment of virtual teaching and research section can also enable teachers and scholars in different fields and disciplines to discuss together around the interdisciplinary knowledge and interdisciplinary elements that the same group of students need to acquire, in order to promote the formation and development of students' interdisciplinary thinking.

4.3 Creating Large-Scale Open and Shared Online National Excellent Courses

Another important aspect of China's engineering education reform is to create excellent courses that can be shared openly, so that students from different schools can benefit from high-quality education and teaching resources through online platforms. This measure is in response to the Department of Higher Education, Ministry of Education's call to develop "Gold Lessons" in recent years, which relies on online course platforms such as Chinese universities' massive open online course (MOOC) platform.

MOOC was first introduced in the United States in 2012, and it quickly gained popularity in China in subsequent years. MOOC is considered an advanced online course that is different from traditional distance learning courses and even general excellent courses. In practical application in China, Chinese universities' MOOC platform generally embodies the following characteristics.

The first feature is the unitization of knowledge, which is based on the discovery of modern brain science and the recognition of learners' new brain laws in the information age. The development of neuroscience and brain science has revealed a fundamental truth about today's students that cannot be ignored: their brains are different from those of just a few years ago, and have adapted to the large amount of stimulation in their environment, filled with myriad distractions and stimuli brought about by technological development and dietary changes (Sousa, 2016, 2017). It is challenging to keep the new generation of students attentive for more than 15 min in a lecture (Pradarelli et al., 2017). Therefore, while MOOCs have a complete curriculum design, their unitized features can ensure that students can improve their cognitive efficiency based on the requirements of fragmented time and visual concentration. On this basis, the content of teaching courses can be continuously optimized and updated, and it is also convenient for the timely integration of different knowledge units.

The second feature of MOOCs is the integration of rich media resources into course contents. One of the basic ideas behind MOOCs is to turn the course into a movie. Unlike traditional teaching, which is presented only with the help of a blackboard and chalk, the process of learning through MOOCs is like following a play or watching a movie, where learners can visually watch and listen to endless color pictures, motions, animation, videos, and narratives in the learning process. Based on the mentioned knowledge points, learners can also see the timely switching of lenses, ensuring that learners have not only visual stimulation but also auditory and even sensory multiple stimulation in learning any knowledge point, thus deepening the construction of the self-knowledge system.

According to the website of the massive open online course platform of Chinese universities (icourse163.org), 810 universities throughout China have created and uploaded their own massive open online courses on this platform. These universities include top comprehensive universities such as Peking University, Tsinghua University, Nanjing University, and Zhejiang University, as well as lesser-known industry-specific universities and local colleges. In this sense, the construction of massive open online courses in China is all-encompassing and extensive, and its participants are not limited to a small number of colleges and universities, but to all colleges and universities at all levels, regions, fields, and industries.

In fact, there are many national projects in China's higher education reform, but many projects, such as double first-class construction, are often limited to dozens of top universities. The creation and development of massive open online courses, however, differs from the participation of these small groups. For this reason, building massive open online courses is regarded as one of the important ways to carry out teaching reform in China's higher education.

At the same time, the massive open online courses offered by these universities and colleges cover all aspects of the engineering field, including mathematics, physics, chemistry, astronomy, geography, biology, electrical information, mechanical engineering, atmospheric and marine agricultural chemistry, civil engineering, hydraulic mechanics, materials, transportation, chemical engineering and biopharmaceuticals, energy, mining, textiles, food, aerospace, agriculture, forestry, and environmental safety.

It is worth noting that all of the MOOC courses on the national platforms are open to students from all universities in China. Clicking on any course on this platform, one can clearly see that the number of participants on average reaches several hundreds to several thousands, and some courses are studied by more than 10,000 learners. For example, the team of Professor Zhu Jianming of the National University of Defense Technology opened an advanced mathematics course on this platform. This course is divided into four phases according to the difficulty of the course content. According to the platform data, the number of students taking each of the four periods has reached more than 10,000 in each phase. Among them, the number of people studying the first period reaches 59,049, the number of learners for Phase 2 and Phase 3 is around 13,000, and the number for Phase 4 is about 25,000. That is to say, there are more than 100,000 people studying Professor Zhu Jianming's Advanced Mathematics throughout the country. Such a large volume is unimaginable if it is only perceived within one university. For this reason, massive open online courses have indeed formed radiation effects in the whole country in a few years in realizing the shared feature of high-quality education and academic resources.

In the process of learning Massive Open Online Courses (MOOCs), students have the opportunity to acquire basic knowledge by watching relevant video teachings at any time, completing exercises corresponding to the course material, and taking online tests. Additionally, MOOC platforms provide students with remote interaction opportunities with teachers and MOOC team assistants. Students can ask professional questions about the course material on the corresponding platform, and teachers or assistants within the course team will provide answers within a specific time frame. Moreover, students can solve their own queries or learn new knowledge by watching the Q&A interaction between other students and the course team. Consequently, students from second-rate and third-rate colleges can access high-quality teaching resources and receive first-class assessment content online.

Furthermore, upon completion of a MOOC, students can obtain a certificate that serves as official recognition of their completion of the course. Some institutions of higher learning accept these certificates as proof of credits earned, which waives the need for students to repeat the same courses in their own universities.

It is important to note that creating high-quality MOOCs requires significant investment in terms of manpower, material resources, time, and finances. To facilitate the creation of quality MOOCs, many universities in China have supported their teaching teams in building MOOCs through special grants.

References

Bridge, P., Appleyard, R. M., Ward, J. W., Philips, R., & Beavis, A. W. (2007). The development and evaluation of a virtual radiotherapy treatment machine using an immersive visualisation environment. *Computers & Education, 49*(2), 481–494.

Chen, J., Huang, Y., Lin, K., Chang, Y., Lin, H., Lin, C., & Hsiao, H. (2020). Developing a hands-on activity using virtual reality to help students learn by doing. *Journal of Computer Assisted Learning, 36*(1), 46–60.

Fan, H. (2022). Speech at the training meeting of pilot construction of virtual teaching and research section. *Smart Teaching Research in Higher Education, 1*(1), 2–3.

Kelley, T. R., & Knowles, J. G. (2016). A conceptual framework for integrated STEM education. *International Journal of STEM Education, 3*(1), 11. https://doi.org/10.1186/s40594-016-0046-z

Laver, J., & Wenger, E. (1991). *Situated learning: Legitimate peripheral participation.* Cambridge University Press.

Lisichenko, R. (2015). Issues surrounding the use of virtual reality in geographic education. *The Geography Teacher, 12*(4), 159–166.

Ostler, T. (1994). Revolution in reality: Virtual reality applications in geography. *Geographical Magazine, 66*(5), 12–13.

Parong, J., & Mayer, R. E. (2018). Learning science in immersive virtual reality. *Journal of Educational Psychology, 110*(6), 785.

Putnam, R. T., & Borko, H. (2000). What do new views of knowledge and thinking have to say about research on teacher learning? *Educational Researcher, 29*(1), 4–15.

Sousa, D. A. (2016). *Engaging the Rewired Brain.* West Palm Beach, FL: Learning Sciences International.

Sousa, D. A. (2017). *How the Brain Learns (5th ed.).* Thousand Oaks: Corwin Press.

Tarng, W., Hsie, C.-C., Lin, C.-M., & Lee, C.-Y. (2017). Development and application of a virtual laboratory for synthesizing and analyzing nanogold particles. *Journal of Computers, 12*(3), 270–283.

Wei, X., Weng, D., Liu, Y., & Wang, Y. (2015). Teaching based on augmented reality for a technical creative design course. *Computers & Education, 81*, 221–234.

Yang, X., Lin, L., Cheng, P.-Y., Yang, X., Ren, Y., & Huang, Y.-M. (2018). Examining creativity through a virtual reality support system. *Educational Technology Research and Development, 66*(5), 1231–1254.

Chapter 5
Fueling Students' Second Classroom Experience Through Disciplinary Contests

Abstract In this chapter, the establishment of the second classroom in Chinese higher education institutions during the reform of engineering education and its instrumental role is elaborated in detail. In contrast to the first classroom, which refers to the educational environment within the classroom setting, the second classroom predominantly denotes a more open and expansive experiential learning environment. In recent years, the contributions of the second classroom in the Chinese engineering education reform encompass dozens of student competitions, both discipline-based and project-based. Compared to the fixed knowledge presented in textbooks and teaching materials, these diverse student competitions are more conducive to stimulating students' expansive thinking, innovative capacities, and applicability skills. This chapter provides a comprehensive overview of the various competitions.

Keywords Second classroom · College student contest · Disciplinary contests · Innovation capacity

The concept of the "second classroom" is coined in the context of Chinese higher education, as opposed to the "first classroom". The "first classroom" refers to classroom teaching activities that take place during the prescribed teaching time, based upon teaching materials and syllabus. In contrast, the "second classroom" refers to teaching activities outside of the first classroom, namely extracurricular activities. The second classroom originates from teaching materials used in formal university settings but is not limited to them in terms of teaching contents. It does not require examinations, but it is an indispensable part of quality-oriented education (Chan, 2016; Marchetti et al., 2016). In terms of form, it is lively and colorful, with a wide learning space that involves a broad range of stakeholders from universities to society (Kholiavko et al., 2020; Sandal et al., 2020). As an important organizational form of student cultivation, the second classroom offers advantages such as being more comprehensive, flexible, and more suitable for students' individual needs. This is helpful in consolidating students' professional knowledge and enhancing their practical skills. When it comes to engineering education, the most important "second

T. Zhuang, *Modernizing China's Undergraduate Engineering Education Through Systemic Reforms*, SpringerBriefs in Education,
https://doi.org/10.1007/978-981-99-6388-1_5

class" that Chinese college students can get in touch with includes disciplinary contests (Xie, 2013; Yan et al., 2018; Zhou & Xu, 2012).

5.1 Versatile Disciplinary Student Contests

In recent years, carrying out versatile disciplinary competitions for college students has become an important means of implementing the teaching idea of the second classroom education. Due to the large number of professional disciplines, the system of student discipline competitions is relatively complex. There are official, folk, community, and school-level competitions at different levels, which can be dazzling. On March 22, 2021, the Research Working Group on College Competition Evaluation and Management System of China Higher Education Association released the 2020 National College Students Competition Rankings. A total of 57 types of contests were listed as officially recognized disciplinary student contests, with the vast majority falling in the area of engineering education.

1.. China "Internet+" College Students Innovation and Entrepreneurship Competition
    https://cy.ncss.org.cn/
2.. "Challenge Cup" National Undergraduate Extracurricular Academic Science and Technology Works Competition
    http://www.tiaozhanbei.net/
3. "Create Youth" Chinese College Students Entrepreneurship Plan Competition
    http://www.chuangqingchun.net/
4. ACM-ICPC International Collegiate Programming Contest
    http://acm.cumt.edu.cn/
5. National Undergraduate Mathematical Contest in Modeling
    http://www.mcm.edu.cn/
6. National Undergraduate Electronic Design Competition
    http://www.nuedcchina.com/
7. National Undergraduate Chemical Experiment Invitational Competition
    http://nuclt.fzu.edu.cn/html/sy/1.html
8. National Medical College Students Clinical Skills Competition
    http://www.kmmc.cn/list1779.aspx
9. National College Students Mechanical Innovation Design Competition
    http://umic.ckcest.cn/
10. National Undergraduate Structural Design Competition
    http://www.ccea.zju.edu.cn/structure/
11. National College Student Advertising Art Competition
    http://www.sun-ada.net/
12. National College Student Smart Car Competition
    https://smartcar.cdstm.cn/index

13. National University Student Transportation Science and Technology Competition
http://www.nactrans.com.cn/index
14. National College Student E-Commerce "Innovation, Creativity and Entrepreneurship" Challenge
http://www.3chuang.net/
15. National College Student Energy Conservation and Emission Reduction Social Practice and Technology Contest
http://www.jienengjianpai.org/Default.asp
16. National College Students Engineering Training Comprehensive Ability Competition
http://www.gcxl.edu.cn/
17. National University Student Logistics Design Competition
http://47.103.191.18/
18. FLTRP National College English Competition—English Speech, English Debate, English Writing, English Reading
http://uchallenge.unipus.cn/
19.. National Vocational College Skills Competition
http://www.chinaskills-jsw.org
20. National University Student Innovation and Entrepreneurship Training Program Annual Meeting Presentation
http://gjcxcy.bjtu.edu.cn/Index.aspx
21. National University Robot Contest—RoboMaster, RoboCon
https://www.robomaster.com/zh-CN
http://www.cnrobocon.net/
22. "Siemens Cup" China Intelligent Manufacturing Challenge
http://www.siemenscup-cimc.org.cn/
23. National Undergraduate Chemical Design Competition
http://iche.zju.edu.cn/
24. National Undergraduate Advanced Graphics Technology and Product Information Modeling Innovation Competition
http://www.dxsgraphics.cn/Default.aspx
25. China Undergraduate Computer Design Competition
http://jsjds.ruc.edu.cn/Index.asp
26. National College Student Market Survey and Analysis Competition
http://www.china-cssc.org/list-52-1.html
27. China University Student Service Outsourcing Innovation and Entrepreneurship Competition
http://www.fwwb.org.cn/
28. "Huacan Award" in the cross-strait cutting-edge design competition
http://www.huacanjiang.com/home
29. China University Computer Competition—Big Data Challenge, Group Programming Ladder Competition, Mobile Application Innovation Competition, Network Technology Challenge
http://www.c4best.cn/

30. World Skills Competition
    http://wscrc.tute.edu.cn
31. World Skills China Trials
32. China Robot Competition and RoboCup Robot World Cup China
    http://crc.drct-caa.org.cn/index.php/race/view?id=663
33. National Undergraduate Information Security Competition
    http://www.ciscn.cn/
34. National Zhou Peiyuan College Students Mechanics Competition
    http://zpy.cstam.org.cn/templates/jiaoyu_001/index.aspx?nodeid=54
35. Chinese College Students Mechanical Engineering Innovation Competition—
    Process Equipment Practice and Innovation Competition, Casting Process
    Design Competition, Material Heat Treatment Innovation and Entrepreneur-
    ship Competition, Crane Creative Competition
    http://cmes-imic.org.cn/
36. Lanqiao Cup National Software and Information Technology Professional
    Talent Competition
    http://dasai.lanqiao.cn/
37. National College Student Metallographic Skills Competition
    http://www.mse-cn.com/
38. "China Software Cup" College Students Software Design Competition
    http://www.cnsoftbei.com/
39. National Undergraduate Photoelectric Design Competition
    http://opt.zju.edu.cn/gdjs/
40. National University Digital Art Design Competition
    http://www.ncda.org.cn/
41. China-US Youth Maker Competition
    http://www.chinaus-maker.org/
42. National Undergraduate Geological Skills Competition
    http://www.saikr.com/33669
43. Milan Design Week-Exhibition of Excellent Works of Teachers and Students
    of Design Disciplines in Chinese Universities
    http://www.hie.edu.cn/announcement_12579/20190627/t20190627_994176.
    shtml
44. National Undergraduate Integrated Circuit Innovation and Entrepreneurship
    Competition
    http://univ.ciciec.com/
45. China Robotics and Artificial Intelligence Competition
    http://www.caai.cn/
46. National University Business Elite Challenge—Brand Planning Competition,
    Exhibition Professional Innovation and Entrepreneurship Practice Compe-
    tition, International Trade Competition, Innovation and Entrepreneurship
    Competition
    http://www.ccpitedu.org/
47. China Good Creativity and National Digital Art Design Competition
    http://www.cdec.org.cn/

48. National 3D Digital Innovation Design Competition
    https://3dds.3ddl.net/
49. "Xuechuang Cup" National College Students Entrepreneurship Comprehensive Simulation Competition
    http://cyds.monilab.com/
50. "Datang Cup" National University Student Mobile Communication 5G Technology Competition
    http://dtcup.dtxiaotangren.com/
51. National Undergraduate Physics Experiment Competition
    http://wlsycx.moocollege.com/
52. National University BIM Graduation Design Innovation Competition
    http://gxbsxs.glodonedu.com/
53. RobCom Robot Developer Contest
    https://www.robocom.com.cn/
54. National Undergraduate Life Science Competition (CULSC)—Life Science Competition, Life Innovation and Entrepreneurship Competition
    https://www.culsc.cn/
55. Huawei ICT Contest
    https://e.huawei.com/cn/talent/#/ict/contest?compId=&navType=talentAlliance
56. National Undergraduate Embedded Chip and System Design Competition
    http://www.socchina.net/
57. China University Intelligent Robot Contest
    http://www.robotcontest.cn/

From the names of the above 57 kinds of college students' discipline competitions, it can be seen that many competitions involve numerous very specific majors, including both basic and applied disciplines. However, most of them are comprehensive discipline competitions sponsored by enterprises or supported by relevant universities, promoting students' comprehensive majors and practical application abilities. The following are some typical engineering-related student contests at the college and university level.

5.2   Typical Examples of Student Contests in Engineering Field

- **China University Robot Competition**

The China University Robot Competition, sponsored by the Central Committee of the Communist Youth League, was held for the first time in 2002. It is a competition for Chinese college students to innovate and start a business in robotics. The competition has always adhered to the purpose of "letting the mind boil and letting wisdom act". The batch of scientific and technological elites who love innovation, know how to get things done, can cooperate, and fight bravely have produced extensive and positive influences in universities and society. The competition currently includes four events: ROBOCON, RoboMaster, ROBOTAC, and the Robot Entrepreneurship Competition. Every year, more than 400 colleges and universities participate, covering nearly 10,000 students from undergraduate and vocational colleges.

ROBOCON is an international college student robot event initiated by the Asia Pacific Broadcasting Federation. China started hosting the event in 2002, and the champion team represented China in the Asia–Pacific University Robotics Contest (ABUROBOCON). The ABUROBOCON competition introduces a new theme every year, and the host country formulates the content and rules of the competition according to its own historical and cultural characteristics. The competition is mainly about engineering tasks, and the participating teams need to complete the complete development process of the robot's design, production, debugging, iteration, and competition within ten months. Students should not only develop robotics, but also form interdisciplinary and interdisciplinary project teams to complete nontechnical tasks such as progress control, finance, materials, and publicity. The competition has played a positive role in cultivating students' comprehensive qualities such as complex engineering cognition, system thinking, teamwork, and project management.

The Chinese representative team (including the Hong Kong Special Administrative Region) has won six championships, seven runner-ups, and six third runner-ups in the ABU annual finals. Among the 100 college student "Xiaoping Science and Technology Innovation Teams" announced by the League Central Committee in 2014, there were 21 robotics-related technology innovation teams. According to statistics, the participating team members have started more than 160 enterprises, about 600 entrepreneurs, and about 16,000 people have been employed. The founders or CTOs of companies such as Dajiang Innovation, Li Qun Automation, Beijing Jizhijia, Yidong Technology, Lexingtianxia, Shenzhen Langchi, Chino Power, Shenzhen Stander, Lingdong Technology, and other enterprises are all participating members of ROBOCON.

- **National College Student Information Security Contest**

The National College Student Information Security Contest is sponsored by the Teaching Guidance Committee for Network Space Security Majors of Higher Education Institutions under the Ministry of Education. It is divided into two categories: Works Competition and Innovative Practice Ability Competition. In 2019, the Works Competition was hosted by Southeast University, and the Innovative Practice Ability Competition was hosted by University of Electronic Science and Technology of China. In 2018, the Works Competition was hosted by Wuhan University, and the Innovative Practice Ability Competition was hosted by Tsinghua University. In 2017, the Works Competition and the Innovative Practice Ability Competition were hosted by Xi'an University of Science and Technology. In 2019, 116 universities across the country participated in the Works Competition, and the Innovative Practice Ability Competition was divided into 8 regions across the country, with 486 universities from 31 provinces, municipalities, and autonomous regions participating.

To select and recommend outstanding professionals in the field of network security, cultivate students' innovation awareness and teamwork skills, improve students' network security technology level, innovation practice, and comprehensive design ability, and promote the construction and reform of network security-related majors in Chinese universities, under the guidance of the Higher Education Department of the Ministry of Education and the Information Security Coordination Bureau of the Central Cyberspace Administration of China, the Higher Education Institution Cybersecurity Instruction Committee has been hosting the National College Student Information Security Contest since 2008, and has successfully held twelve contests to date. The competition is open to full-time undergraduate and college students with formal enrollment throughout the country. The contest is divided into two categories: the works competition focuses on the development of network security systems, using open-ended topics and independent design; the innovation practice ability competition focuses on practical network security attack and defense, aiming to improve students' innovation and practical skills in the field of network security.

- **National Zhou Peiyuan Mechanics Competition for College Students**

The National Zhou Peiyuan Mechanics Competition for College Students is jointly organized by the Higher Education Mechanics Foundation Course Teaching Guidance Subcommittee of the Ministry of Education, the Chinese Society of Mechanics, and the Zhou Peiyuan Foundation. In 2019, the competition was hosted by the Editorial Committee of Mechanics and Practice and the Education Work Committee of the Chinese Society of Mechanics. The first competition was held in 1988 and has been held every 2–3 years since then. As of CMCZ, there have been 12 competitions, covering 30 provinces, municipalities, and autonomous regions throughout the country. The number of applicants has increased with each competition, and the total number of participating students has exceeded 100,000 in the past 5 competitions. In 2019, there were a total of 30 competition areas, with over 400 undergraduate colleges and universities participating from 30 provinces, municipalities, and autonomous regions.

The National Zhou Pei-yuan Mechanics Competition for College Students is a science and technology event commissioned by the Ministry of Education, aimed at serving education and nurturing talent. It is a competition to promote the reform of mechanics fundamental courses in higher education and to increase students' interest in learning basic mechanics. It is also a competition to strengthen the quality education of science and engineering students, cultivate their hands-on ability, innovation ability, and team spirit. Moreover, it is an extracurricular activity that tests whether young students can flexibly apply classroom mechanics knowledge, and discovers and selects innovative talents for the future. The competition is open to undergraduate, college, and graduate students. It includes both individual and team competitions. The individual competition adopts a closed-book written examination, and theoretical mechanics and material mechanics are combined into one test paper. The team competition is divided into "theoretical design and operation" and "basic mechanics experiments," and adopts a group project research (experimental testing) approach.

- **The "Blue Bridge Cup" National Software and Information Technology Professionals Competition**

The "Blue Bridge Cup" National Software and Information Technology Professionals Competition is a national competition for software and information technology professionals in China. It was initiated in 2010 by the Ministry of Education, the Ministry of Industry and Information Technology, the Chinese Computer Federation, and the People's Government of Shandong Province. The competition is held annually and consists of two categories: software engineering and information technology. It is open to college and vocational school students across the country. The competition aims to improve the skills and innovation ability of software and information technology professionals, promote the development of the software and information technology industry, and enhance the competitiveness of Chinese software and information technology talents on the international stage.

As of 2020, the National Blue Bridge Cup has been held for 11 consecutive years, with a total of more than 300,000 participants. In 2019, 1191 colleges and universities from 31 provinces, municipalities, and autonomous regions participated, with a total of 60,746 participants. The competition has always adhered to the purpose of "focusing on the industry, highlighting practice, broad participation, and promoting employment", focusing on the key areas of information technology that are urgently needed for social development, and promoting the development of T-professional skills and innovative abilities of college students.

The competition subjects of the Blue Bridge Cup include C/C++ programming, Java software development, Python programming, single-chip microcontroller design and development, embedded design and development, Internet of Things design and development, graphic design, animation design and production, video design and production, etc., covering the current mainstream information technology professional skills. The competition topics of the Blue Bridge Cup are characterized by engineering, skills, and fun. The topic setting is related to important links or common problems in practical engineering applications. The solution to the problem has reference significance for actual development. At the same time, the competition topic has a novel idea, and the algorithm is reasonably applied, even including mathematical results. The topic emphasizes fun, inspiring thinking and making learning enjoyable through competition topics.

The software and electronic categories of the competition are conducted on an individual basis. Tens of thousands of participants complete the independent closed-book test at designated competition points, which truly tests the abilities and technical level of the students. The Blue Bridge Cup fully affirms the subject status of students' learning, mobilizes students' subjective initiative in learning, and has a significant promoting effect on cultivating students' practical and innovative abilities, and improving the practical teaching mode and innovative talent training mode of universities.

As an IT competition for college students across the country, over 1000 colleges and universities actively participate, and a large number of competition participants have entered first-class enterprises such as IBM, Microsoft, Huawei, Alibaba, etc.

At the same time, the Blue Bridge Cup has gone abroad, holding three international competitions in the United States and Germany, and setting up a branch competition in Japan, attracting students from foreign universities such as MIT, Berlin Institute of Technology, and the University of Tokyo to participate, becoming an important platform for young people from China and foreign countries to enhance mutual understanding and deepen mutual feelings.

- **Chinese College Students Mechanical Engineering Innovation and Creativity Competition**

The MEICC (Mechanical Engineering Innovation and Creativity Competition for University Students) is organized by the China Mechanical Engineering Society. With the guidance of the Higher Education Teaching Guidance Committee of the Ministry of Education, the competition aims to promote innovation and creativity among university students majoring in mechanical engineering, materials science, industrial engineering, and related fields. It encourages students to keep up with the forefront of science and technology, participate in engineering and technological innovation, and cultivate teamwork and craftsmanship. The competition serves to enhance students' engineering practical skills and innovation capabilities, contribute to the national strategy of building a manufacturing power, and accelerate the cultivation of innovative and entrepreneurial talents in the equipment manufacturing industry.

The competition includes seven professional events, namely "Process Equipment Practice and Innovation Competition", "Casting Process Design Competition", "Material Heat Treatment Innovation and Entrepreneurship Competition", "Crane Creative Competition", "Industrial Engineering and Lean Management Innovation Competition", "Intelligent Manufacturing Competition", and "Micro-Nano Sensor Technology and Intelligent Application Competition". In 2019, more than 10,000 students from 2305 teams representing 523 universities across the country participated in the competition.

The "Process Equipment Practice and Innovation Competition" consists of three stages: registration and preparation of works, communication evaluation by experts, and results reporting and on-site evaluation. After the submission of the works, the expert committee organizes anonymous communication evaluation to select outstanding works for the competition. The selected projects enter the final round of the competition for on-site evaluation and defense. The competition schedule is announced every October, team registration is opened in April of the following year, and the submission of works takes place in June. The preliminary online evaluation is completed by the end of July, and the outstanding works are selected. The exhibition and final round of the competition are held in mid to late August.

The "Casting Process Design Competition" includes school-level preliminary, semi-final, and final rounds. The participating schools select works for the semi-final round through the preliminary round, and the expert committee evaluates the semi-final works anonymously to determine the works for the final round. The works in the final round are evaluated and ranked through on-site defense. The competition schedule is announced every October, team registration is opened in April of the following year, and the submission of works takes place in June. The preliminary online evaluation is completed by the end of July, and the outstanding works are selected. The exhibition and final round of the competition are held in mid to late August.

The "Material Heat Treatment Innovation and Entrepreneurship Competition" consists of two parts: the preliminary round and the final round. The preliminary round includes optional project speeches and displays, and basic knowledge quizzes. The final round includes the comprehensive display and application of material heat treatment related knowledge. The application period is in March, the submission of works is in April and May, the formal review and public announcement takes place in June, and the final round is held from July to October.

The "Crane Creative Competition" is divided into a school-level selection round and a national final. Each participating school selects no more than five works for the national final through its own selection process. The works with foreign students' participation are not limited by the number of works selected by each school. The final works are evaluated and ranked by the judging panel according to the actual completion and design novelty of the competition rules. The competition notice is officially released before the winter vacation every year. The participating schools register in late May, and organize the school-level selection round by early August. The national final is held from September to October.

Overall, the MEICC is an annual competition that aims to cultivate innovative and practical talents in the field of mechanical engineering and related subjects, to promote scientific and technological progress, and to contribute to the development of the country's manufacturing.

- **China Software Cup" National College Student Software Design Competition**

The "China Software Cup" National College Student Software Design Competition is an annual software design competition for college students in China. It is organized by the Ministry of Education, the Ministry of Industry and Information Technology, and the Chinese Institute of Electronics, and is one of the largest and most prestigious software competitions in China. The competition aims to encourage innovation, creativity, and teamwork among college students, and to promote the development of the software industry in China. Participants are typically required to design and develop software solutions for real-world problems or challenges, and the entries are judged by a panel of experts from industry, academia, and government. Winners of the competition often receive recognition, prizes, and opportunities to further develop their ideas and projects. Many successful software companies and entrepreneurs in China have participated in and won the China Software Cup competition, making it an important platform for talent development and innovation in the country's software industry.

Since its launch in 2011, the "China Software Cup" National College Student Software Design Competition has been held for 8 consecutive years. In 2019, 5524 teams from 809 universities (including vocational schools) in 31 provinces, municipalities, and autonomous regions participated in the competition, with more than

20,000 teachers and students taking part. Over the past nine years, the competition has always adhered to the principle of "government guidance, enterprise problem setting, university participation, expert evaluation, talent nurturing and selection", covering every aspect of software talent development. With the competition as the driving force, the innovation mechanism and collaboration form of software talent cultivation have been strengthened, and a new model of software talent cultivation with "government-industry-university-research-application" as the core has been created.

The competition has deepened cooperation between universities and enterprises, ensured talent output, and formed a talent "think tank" for backbone enterprises in the software and information service industry through various activities such as talent recruitment fairs and employment intention agreements, using the competition as a carrier. It has explored a path of interactive production and education, win–win cooperation, software talent cultivation and output practice. The competition has effectively brought together the strength of both enterprises and universities to enhance the collaborative innovation ability of production, education and research, aiming to jointly support the construction of a safe and reliable system. It encourages independent software core enterprises to rely on the competition platform to realize technology introduction, talent cultivation and brand promotion, and provides guidance and support for the development direction of independent software technology application in the review of competition topics. It emphasizes on cultivating the ability of university students in the development and application of domestic software.

Since its inception, the competition has gathered hundreds of domestic software backbone enterprises, including over 1000 undergraduate and vocational colleges and universities, including all 985 and 211 universities. It has attracted tens of thousands of teams and more than 200,000 software and computer professionals have participated in the competition, helping the top 100 software enterprises in China to solve nearly 300 common technical problems. The competition topics closely follow the hot spots of industry development and focus on solving enterprise pain points. In terms of organization, the competition emphasizes on "promoting education and learning through competition" to make the participation benefits sustainable for teachers and students. Each competition lasts for one year, usually starting in September and ending with the national finals in September of the following year. The main stages include: collection and revision of competition topics, topic release, registration organization, online coaching, work submission, preliminary evaluation, regional and special finals, and national finals (simultaneously launching the next competition).

- **National College Student Optoelectronic Design Competition**

The National College Student Optoelectronic Design Competition is an annual event held in China to promote innovation and development in the field of optoelectronics. The competition is open to all college students across the country and aims to encourage the application of optoelectronic technology in various fields such as energy, healthcare, and communication. The competition is typically held in several stages, beginning with regional contests and culminating in a national final. Participants are required to design and develop an optoelectronic system or product that meets certain criteria, such as cost-effectiveness, energy efficiency, and practicality.

The competition is organized by the Ministry of Education, the Ministry of Industry and Information Technology, and the Chinese Optical Society. The organizers provide support and guidance to the participants throughout the competition, and winners receive recognition and awards for their achievements. The National College Student Optoelectronic Design Competition has played an important role in fostering the development of the optoelectronics industry in China. It has encouraged innovation and collaboration among students, universities, and businesses, and has helped to identify and develop new talent in this field. Additionally, the competition has helped to raise awareness of the potential applications of optoelectronic technology and promote its adoption in various industries.

The 2019 National Finals of the National College Student Optoelectronic Design Competition were held in Qingdao, with a total of 179 undergraduate colleges and universities from 29 provinces, municipalities, and autonomous regions participating in 7 regional competitions. The finals of the seven regional competitions, including North China, Northeast China, Northwest China, Southwest China, Central China, Southeast China, and East China, were held respectively by Taiyuan University of Technology, Lianyungang University, Xi'an University of Posts and Telecommunications, Chongqing University of Posts and Telecommunications, Nanchang Hangkong University, Quanzhou Normal University, and Nanjing University of Science and Technology. The competition is open to undergraduate and graduate students, as well as international students, and encourages cross-school, cross-disciplinary, and cross-major team participation, with each team consisting of at least 2 undergraduate students, and each student participating in only one team. The competition is mainly aimed at students majoring in optoelectronics from colleges and universities across the country, but also provides opportunities for students from various disciplines to learn and apply optoelectronics knowledge to solve practical problems, stimulate creative thinking, and appreciate the charm of optoelectronics. The competition aims to promote education equity and strengthen the cross-fusion of optical engineering disciplines with other disciplines. Since its establishment in 2008, the competition has been held seven times. Each of the first six competitions had two themes, namely "Light and Energy," "Light and Life," "Light and Information," "Light and Measurement," "Intelligent Light," and "Exploration of Light." The seventh competition had the theme "Standing at the Forefront of the Times, Displaying the Charm of Optoelectronics," and held the first optoelectronic specialty creative competition. The National Finals of the seven competitions were held respectively by Zhejiang University, Changchun University of Science and Technology, Fujian Normal University, National University of Defense Technology, University of Electronic Science and Technology of China, Beijing Institute of Technology, and Qingdao University, covering the north, south, east, and west regions of China, fully leveraging the regional radiation-driven effect. There are about 262 colleges and universities in China with optoelectronic majors, and more than two-thirds of them have signed up for the National College Student Optoelectronic Design Competition. The competition has attracted a total of 595 universities and 13,544 student participants, with optoelectronic companies such as Omron Automation (China) Co., Ltd., Hamamatsu Photonics Trading (China) Co., Ltd., Sunlord Group, and Goertek Group participating in sponsorship, and attracting the attention of many media outlets, including Guangming Daily, Science and Technology Daily (China Science and Technology Network), and the Central Committee of the Communist Youth League News Center (China Youth Network).

- **National College Students' Integrated Circuit Innovation and Entrepreneurship Competition**

The National College Students' Integrated Circuit Innovation and Entrepreneurship Competition, also known as the "Circuits Cup" in China, is an annual competition that aims to promote innovation and entrepreneurship in the field of integrated circuits among college students in China. The competition was first held in 2012 and has since grown in size and importance. The competition is organized by the Ministry of Education, the Ministry of Industry and Information Technology, the Ministry of Science and Technology, and the China Association for Science and Technology. It is also supported by various industry associations, universities, and companies in the field of integrated circuits. The competition consists of three rounds: the preliminary round, the semi-final round, and the final round. In the preliminary round, students submit their project proposals, which are evaluated based on their innovation, feasibility, and commercial potential. The semi-final round involves the development of a prototype, and the final round includes a presentation of the project to a panel of judges, which typically consists of industry experts, investors, and academics.

The competition is open to undergraduate and graduate students from universities across China, and students are encouraged to form teams with members from different academic backgrounds and disciplines. The competition covers a wide range of topics in the field of integrated circuits, including but not limited to analog circuits, digital circuits, mixed-signal circuits, system-on-chip (SoC) design, and semiconductor manufacturing. The competition has gained significant attention and recognition in recent years. In 2019, the competition attracted over 13,000 participants from 520 universities across China. The winning teams are awarded with prizes and funding opportunities to support the commercialization of their projects. The competition has also been recognized as a platform for promoting innovation and entrepreneurship in the field of integrated circuits and for fostering collaboration between academia and industry.

Overall, the versatile college students' disciplinary contests play an important role in promoting their second classroom experiences in several ways. It encourages interdisciplinary collaboration in that disciplinary contests often require students from different academic disciplines to work together to solve a problem or complete a project. This encourages interdisciplinary collaboration and helps students develop a more comprehensive understanding of the subject matter. At the same time, it fosters creativity and innovation by challenging students to think creatively and come up with innovative solutions to problems. This helps them develop their problem-solving skills and fosters a spirit of creativity and innovation. Furthermore, it provides real-world experience by enabling participants to have real-world experience that can be applied to their future careers. This hands-on experience helps students develop practical skills and gives them a better understanding of how their academic knowledge can be applied in real-world situations. Moreover, it builds confidence and leadership skills as students who participate in disciplinary contests often gain confidence in their abilities and develop leadership skills as they work with their peers to achieve a common goal. This can be particularly valuable for students who may not have had opportunities to develop these skills in the classroom. Overwhelmingly, disciplinary contests can help promote a more well-rounded and enriching educational experience for college students by providing opportunities for interdisciplinary collaboration, fostering creativity and innovation, providing real-world experience, and building confidence and leadership skills.

References

Chan, Y.-K. (2016). Investigating the relationship among extracurricular activities, learning approach and academic outcomes: A case study. *Active Learning in Higher Education, 17*(3), 223–233.

Kholiavko, N., Detsiuk, T., & Tarasenko, O. (2020). Extracurricular activity of engineering students: Trends and motives. *Journal of Educational Sciences & Psychology, 10*(1).

Marchetti, R., Wilson, R. H., & Dunham, M. (2016). Academic achievement and extracurricular school activities of at-risk high school students. *Educational Research Quarterly, 39*(4), 3–20.

Sandal, J.-U., Detsiuk, T., & Kholiavko, N. (2020). Developing foreign language communicative competence of engineering students within university extracurricular activities. *Advanced Education, 14*, 19–28.

Xie, J. (2013). An introduction to CUMCM: China/contemporary undergraduate mathematical contest in modeling. In *Educational interfaces between mathematics and industry: Report on an ICMI-ICIAM-study* (pp. 435–443).

Yan, X., Gu, D., Liang, C., Zhao, S., & Lu, W. (2018). Fostering sustainable entrepreneurs: Evidence from China college students' "Internet Plus" innovation and entrepreneurship competition (CSIPC). *Sustainability, 10*(9), 3335.

Zhou, M., & Xu, H. (2012). A review of entrepreneurship education for college students in China. *Administrative Sciences, 2*(1), 82–98.

Chapter 6
Establishing Modern Industrial Colleges and Schools of Future Technology

Abstract This chapter elaborates upon the two structural arrangements of China's engineering education reform, namely the establishment of Modern Industrial Colleges and Schools of Future Technology. Whereas the former prioritizes the application of current information technology and resource innovation to enhance teaching and education quality, the latter focuses on cutting-edge, revolutionary, and disruptive technologies projected to emerge within the next 10–15 years. Their objective is to transcend traditional conventions, constraints, and barriers while fostering change, innovation, and leadership. These institutions aim to cultivate forward-looking, pioneering technology innovators who can spearhead future development and contribute to the pursuit of high-quality development in building a robust higher education nation.

Keywords Modern industrial colleges · School of future technology · Institutional reform to engineering education

In addition to implementing pedagogical transformations within conventional learning environments and incorporating industrial influence into traditional classrooms, another comprehensive approach to higher engineering education reform in China involves the establishment of modern industrial colleges intimately linked with industries and enterprises. Employing a metaphor, the integration of industrial forces into conventional classroom instruction to effectuate teaching reform may be likened to adjusting a machine; however, the founding of a modern industrial college represents the systematic placement of this apparatus within a holistic industrial context for comprehensive immersion. In 2020, two crucial ministries, the Ministry of Education and the Ministry of Industry and Information Technology, collaboratively issued a directive concerning the creation of modern industrial colleges (henceforth referred to as the 'Guideline'), with the aim of executing the State Council's viewpoints on the amalgamation of the industrial and higher education sectors, as well as the explicit criteria for cultivating exceptional engineers.

In contrast, Schools of Future Technology place greater emphasis on the vanguard of industries and technologies. Unlike modern industrial colleges, which prioritize

T. Zhuang, *Modernizing China's Undergraduate Engineering Education Through Systemic Reforms*, SpringerBriefs in Education, https://doi.org/10.1007/978-981-99-6388-1_6

the application of current information technology and resource innovation to enhance teaching and education quality, Schools of Future Technology focus on cutting-edge, revolutionary, and disruptive technologies projected to emerge within the next 10–15 years. Their objective is to transcend traditional conventions, constraints, and barriers while fostering change, innovation, and leadership. These institutions aim to cultivate forward-looking, pioneering technology innovators who can spearhead future development, facilitate the transition from "Made in China" to "Innovation in China," and contribute to the pursuit of high-quality development in building a robust higher education nation. This approach is also considered a top-level strategy for implementing and promoting the comprehensive enhancement of engineering education quality in China.

6.1 The Overarching Framework for Modern Industrial Colleges Development

The Guideline anticipates that higher education institutions should orient themselves towards alignment with local economic and societal development, continuously optimize their professional structures, enhance their operational vitality, explore effective linkage mechanisms among industrial, innovation, and education chains, establish a novel information-sharing mechanism for talent, technology, and material resources, improve the integrated production-education collaborative mechanism, innovate evaluation and employment mechanisms for enterprise part-time instructors, construct a linkage development mechanism between higher education and industrial clusters, and create a cadre of demonstrative talent that amalgamates functions such as talent training, scientific research, technological innovation, enterprise service, and student entrepreneurship. Specifically, the establishment of modern industrial colleges entails seven primary tasks.

The first task is to innovate the talent training model. The Guideline mandates that modern industrial colleges facilitate open collaboration among multiple agents, integrate diverse innovation components and resources, and distill an applied talent training model that deeply merges production and education, as well as multi-party cooperative education.

The second task is to enhance the quality of engineering programs. Modern industrial colleges should concentrate on key development areas designated by state and local governments, endeavor to foster the integration and advancement of new engineering, new agricultural science, new medical science, and new liberal arts, deepen the construction of professional connotations, actively adjust the professional structure, focus on cultivating specialties with distinctive features and advantages, and promote the clustered development of specialties. They must establish close connections with the industrial chain, actualize multidisciplinary cross-collaboration, and support the rapid development of several related majors within the same industrial chain. In accordance with the frontier trends of industry and industrial development, these institutions should advance the establishment of several new applied undergraduate majors and explore innovative development pathways for undergraduate majors. Moreover, they should promote cooperation with enterprises to form a professional construction steering committee, introduce industry standards and enterprise resources to actively pursue international substantive equivalent professional certification, enhance coordination and linkage between professional certification and entrepreneurial and employment qualifications, and elevate the standardization and internationalization levels of program development.

The third task is to develop courses rooted in university-industry collaboration. Enterprises are encouraged to actively participate in textbook development and course construction, design course systems, optimize course structures, accelerate the renewal of course teaching content, attend to the dynamic progression of industry innovation chains, foster the scientific integration of course content with industry standards, production processes, project development, and other industry requirements, and create a collection of high-quality cooperative courses, textbooks, and engineering case sets between schools and enterprises. Building upon the technological innovation projects of industry enterprises and closely integrating actual industrial innovation teaching content, methods, and tools, the proportion of comprehensive and design-oriented practical teaching should be increased. Real projects and product designs of industry enterprises should serve as the basis for topic selection in graduation design, course design, and other practical components. Grounded in professional characteristics, immersive, hands-on, and on-site instruction should be conducted using real production lines and other environments, with an emphasis on enhancing students' practical skills, effectively improving their understanding of the industry and their ability to tackle complex problems.

The fourth task is to establish high-quality internship and training bases. Innovating the cooperation mode among multiple entities, practical teaching and training environments should be constructed based on the products, technologies, and production processes of industry enterprises, as well as industry development and innovation needs. Various practical teaching resources should be coordinated, fully utilizing high-quality resources such as technology industrial parks and industry-leading enterprises, and building professional or cross-disciplinary practical teaching platforms characterized by intensive functionality, open sharing, and efficient operation. By incorporating enterprise research and development platforms and production bases, a number of large-scale experimental and practical internship bases that integrate production, teaching, research and development, and innovation and entrepreneurship functions should be developed, fostering collaboration among industry, academia, research, and application sectors.

The fifth task is to construct a high-level teaching staff. The Guideline mandates reliance on modern industrial colleges, exploration of a two-way talent flow mechanism between educational institutions and enterprises, the establishment of flexible personnel systems, and the development of effective pathways for selecting industry associations, enterprise operational backbones, and exceptional technical and management talents to teach at universities. The implementation of a special position plan for industrial teachers (mentors) should be explored, and mechanisms for introducing, certifying, and utilizing part-time industrial teachers should be improved. Teacher training should be strengthened, fostering collaboration in the creation of practical positions for teacher enterprises, carrying out teacher exchanges, discussions, training, and other businesses, and transforming the modern industrial college into a "dual teacher and dual ability" teacher training base. Joint teaching and guidance between school and enterprise mentors should be conducted, teacher incentive systems should be explored, and a high-level teaching team should be established.

The sixth task is to construct a service platform and nexus for industry-education-research collaboration. Universities and enterprises are encouraged to integrate resources from both parties, establish joint laboratories (research and development centers), leverage the comprehensive advantages of the university's talent and professional expertise, undertake collaborative innovation surrounding key issues of industrial technology innovation, facilitate knowledge spillover from universities to directly serve regional economic and societal development, promote the transformation and application of applied research findings, and advance industrial transformation and upgrading. The joint efforts of schools and enterprises in technology research, product development, achievement transformation, and project incubation should be reinforced, collaboratively completing teaching and research tasks, sharing research outcomes, generating a collection of scientific and technological innovation accomplishments, and enhancing the competitiveness of industrial innovation development. The integration of science and education should be vigorously promoted, research results should be promptly introduced into the teaching process, active interaction between scientific research and talent cultivation should be encouraged, the demonstration influence of industry-university research collaboration should be leveraged, and service industry capacity should be augmented.

The seventh task is to improve the management mechanism of all stakeholders to jointly advance the reform and development of higher engineering education. The Guideline emphasizes the need to strengthen the collaboration among multiple entities such as universities, local governments, industry associations, and enterprise institutions, to create an organizational structure of joint construction and management, explore governance models such as councils and management committees, endow modern industrial colleges with the requisite human rights, administrative and financial privileges for reform, and construct a scientific, efficient, and robust institutional system. Thorough consideration should be given to the characteristics of regions, industries, and sectors, combined with the inherent endowment characteristics of universities, in order to optimize the allocation mode of innovative resources, enhance the "self-hematopoietic" capacity, and create a demonstration zone for the integration of industry and education in universities, ultimately achieving deep integration of the education chain, innovation chain, and industry chain.

In 2021, the Ministry of Education announced the first batch of 50 modern industrial colleges in China, each of which is affiliated to a higher education institution. In 2023, the Ministry of Education has just started the application procedures of the second batch of modern industrial colleges and the work is still in progress. Table 6.1 shows the detailed list of the first batch of modern industrial colleges and the higher education institutions to which they are affiliated.

Table 6.1 List of the first batch of modern industrial colleges

No.	Name of the modern industrial college	Affiliated to	Province (region, city)
1	College of Modern Industry of Traditional Chinese Medicine and Pharmacy	Tianjin University of Chinese Medicine	Tianjin
2	Intelligent Automotive Industry College	Hebei University of Technology	Hebei Province
3	Wine Institute	Hebei Normal University of Science and Technology	Hebei Province
4	School of Information and Innovation Industry	North University of China	Shanxi Province
5	Institute for Tourism Studies	Inner Mongolia Normal University	Inner Mongolia Autonomous Region
6	Magnesite Industrial College	Shenyang University of Chemical Technology	Liaoning Province
7	CRRC Academy	Dalian Jiaotong University	Liaoning Province
8	Big Data Industry College	Bohai University	Liaoning Province
9	Yatai Digital Construction Industry College	Jilin Jianzhu University	Jilin Province
10	Institute of Modern Industry of Authentic Medicinal Materials	Jilin Agricultural University	Jilin Province
11	Beidahuang Agricultural Products Processing Modern Industry College	Heilongjiang Bayi Agricultural Reclamation University	Heilongjiang Province
12	Joint College of Modern Biomedical Industry	East China University of Science and Technology	Shanghai
13	School of Modern Industry of New Materials	Donghua University	Shanghai
14	Shanghai Institute of Microelectronics Industry	Shanghai University	Shanghai
15	2011 Membrane Industry College	Nanjing University of Technology	Jiangsu Province
16	Alibaba Cloud Big Data Academy	Changzhou University	Jiangsu Province
17	School of Artificial Intelligence and Intelligent Manufacturing	Jiangsu University	Jiangsu Province
18	Artificial Intelligence Industry College	Nanjing University of Information Science and Technology	Jiangsu Province
19	Tongke School of Microelectronics	Nantong University	Jiangsu Province
20	New Energy Academy	Yancheng Institute of Technology	Jiangsu Province
21	Nari Institute of Electrical and Automation	Nanjing Normal University	Jiangsu Province

(continued)

Table 6.1 (continued)

No.	Name of the modern industrial college	Affiliated to	Province (region, city)
22	School of Photovoltaic Technology	Changshu Institute of Technology	Jiangsu Province
23	Intelligent Manufacturing Industry College	Changzhou Institute of Technology	Jiangsu Province
24	Intelligent Manufacturing Equipment Industry College	Yangzhou University	Jiangsu Province
25	Digital Manufacturing Industry College	Zhejiang University of Technology	Zhejiang Province
26	Hangzhou Bay Automotive College	Ningbo Institute of Technology	Zhejiang Province
27	Intelligent Manufacturing Modern Industry College	Hefei University of Technology	Anhui Province
28	Robot Modern Industry College	Anhui Polytechnic University	Anhui Province
29	Intelligent Manufacturing Industry College	Fujian Institute of Technology	Fujian Province
30	Advanced Copper Industry Institute	Jiangxi University of Science and Technology	Jiangxi Province
31	Intelligent Equipment Manufacturing Industry College	Henan University of Science and Technology	Henan Province
32	Chip Industry Academy	Hubei University of Technology	Hubei Province
33	Dongfeng HUAT Intelligent Automobile Industry College	Hubei Institute of Automotive Technology	Hubei Province
34	Rail Transit Modern Industry College	Central	Hunan Province
35	Golden Mile Inspection Institute	Guangzhou Medical University	Guangdong Province
36	Tencent Cloud Artificial Intelligence Academy	Shenzhen University	Guangdong Province
37	Intelligent Software Academy	Guangzhou University	Guangdong Province
38	Guangdong-Hong Kong Robotics College	Dongguan Institute of Technology	Guangdong Province
39	Siemens Intelligent Manufacturing Academy	Dongguan Institute of Technology	Guangdong Province
40	School of Integrated Circuit Design Industry	Guangdong University of Technology	Guangdong Province
41	School of Semiconductor Optical Engineering Industry	Foshan Institute of Science and Technology	Guangdong Province
42	Intelligent Vehicle (Manufacturing) and New Energy Vehicle Industry College	Guangxi University of Science and Technology	Guangxi Zhuang Autonomous Region

(continued)

Table 6.1 (continued)

No.	Name of the modern industrial college	Affiliated to	Province (region, city)
43	Industrial Internet Academy	Chongqing University of Posts and Telecommunications	Chongqing
44	New Energy Vehicle Modern Industry College	Chongqing University of Technology	Chongqing
45	CRRC Times Institute of Microelectronics	Southwest Jiaotong University	Sichuan Province
46	College of Modern Natural Gas Industry	Southwest Petroleum University	Sichuan Province
47	College of Modern Industry of Health and Medicine	Guizhou Medical University	Guizhou Province
48	Artificial Intelligence Industry College	Kunming University of Science and Technology	Yunnan Province
49	School of Modern Wine Industry	Northwest A&F University	Shaanxi Province
50	Intelligent Manufacturing Modern Industry College	Xinjiang University	Xinjiang Uygur Autonomous Region

6.2   Piloting Schools of Future Technology in Top-Notch Universities

In December 2021, the Ministry of Education announced the inaugural batch of 12 Schools of Future Technology in China. The Ministry mandated that all Schools of Future Technology excel in three key areas: setting development objectives, pursuing diversified exploration, and enhancing quality assurance.

Regarding the definition of development objectives, the Ministry stipulates that all Schools of Future Technology must target cutting-edge, revolutionary, and disruptive technological breakthroughs, as well as conventional and constraint-breaking advancements within the next 10–15 years. These institutions must strive to nurture forward-looking technological innovation leaders capable of guiding future development, thereby establishing a foundation for constructing a powerful higher education nation and facilitating high-quality development.

Concerning diversified exploration, the Ministry requires relevant colleges and universities establishing Schools of Future Technology to harmonize the direction of personnel training with the institutions' specific contexts to distill the unique features of future technologies. These colleges must undertake varied explorations of School of Future Technology construction models, starting with the provision of high-quality resources and fostering a conducive atmosphere. The Ministry encourages these colleges to establish multidisciplinary, interdisciplinary mechanisms, investigate diverse training modalities, design multi-exit systems, foster cross-disciplinary collaborative education, promote comprehensive multi-faceted changes, and create a teaching and research stronghold capable of guiding future technological development and cultivating innovative leaders.

As for strengthening quality assurance, the Ministry mandates that all relevant universities improve the quality assurance systems of their Schools of Future Technology, thoroughly implement student-centered, output-oriented, and continuously improving concepts, and integrate quality awareness, standards, evaluation, and management throughout the process of fostering future technological innovation leaders. The Ministry requires all Schools of Future Technology to establish a student growth database, continuously refine training programs and processes, coordinate resources, support and prioritize college construction through policies and funding, and promote stable development.

According to pertinent information from the Ministry of Education, the establishment of Schools of Future Technology represents an advanced iteration of the current new engineering education paradigm. These institutions primarily concentrate on two crucial objectives: first, cultivating preeminent talent in key core technologies for the future, and second, investigating and advancing these vital core technologies. With regard to emphasizing future core technologies, the Ministry of Education specifically envisions these Schools of Future Technology addressing critical bottlenecks in key areas through fundamental research over the next several years, thus amassing forward-looking technological reserves to foster the nation's enduring innovation capacity.

Consequently, universities are well-positioned to establish Schools of Future Technology based on their preexisting dominant disciplines. These colleges primarily focus on three categories. The first is the interdisciplinary, cross-integrated domain, which constitutes the central emphasis of each future technical college. For instance, Northeastern University's Future Technical College centers on industry and the intelligent domain, nurturing leading talents in scientific and technological innovation who will spearhead the evolution of future industrial intelligent technology through the cross-integration of premier disciplines and majors, such as control science and engineering, computer science and technology, software engineering, and robotics science and engineering. Peking University's School of Future Technology, on the other hand, concentrates on future life and health technology, delving into two cutting-edge interdisciplinary subjects: biomedical engineering and molecular medicine.

The second category of focus for the Schools of Future Technology encompasses various strategic areas highly valued by nations worldwide. For example, Beihang University's Future Institute of Aerospace Technology targets the frontier fields of aerospace science and technology, aiming to educate a cadre of future aerospace system masters endowed with imagination, insight, executive leadership, and other core qualities. This education spans the entirety of the institution, including information majors, aerospace majors, national first-class undergraduate major construction sites, and their doctoral stage.

Lastly, the third category encompasses emerging technology fields, such as quantum technology, which is the primary focus of the University of Science and Technology of China's Institute of Future Technology.

In comparison to modern industrial colleges, Schools of Future Technology necessitate a greater emphasis on the integration of comprehensive capabilities, technical expertise, and educational prowess. Consequently, the inaugural cohort of pilot colleges is relatively small, consisting of only 12 institutions. These 12 establishments represent the crème de la crème of Chinese higher education, including Peking University, Tsinghua University, Beihang University, Tianjin University, Northeastern University, Harbin Institute of Technology, Shanghai Jiao Tong University, Southeast University, University of Science and Technology of China, Huazhong University of Science and Technology, South China University of Technology, and Xi'an Jiao Tong University.

Consequently, the application for a school of future technology demands heightened prerequisites. The institutions submitting proposals must base their submissions on content derived from nationally recognized first-class majors or first-class disciplines, capitalizing on their unique strengths. These institutions must also possess a stable, high-caliber teaching staff, as well as substantial educational and research resources. Furthermore, the proposing colleges and universities must establish an advanced management system characterized by innovative ideas and efficient operations, while also providing ample, centralized physical space. In addition, these institutions must ensure consistent financial support for staff appointments and the day-to-day operations of the schools. And it is essential to provide the necessary policy backing to foster the development of future technical colleges.

Chapter 7
Carrying Out Engineering Education Research at All Levels

Abstract Compared to the 1990s, when engineering education was only a small branch in the overall landscape of higher education research, today, engineering education research has taken up a prominent place in Chinese higher education research, especially since the beginning of the twenty-first century, representing approximately half of the total research output. In this chapter, a detailed examination of the widespread emergence of engineering education research in China is provided, identifying seventeen critical areas highlighted in key policy documents related to the New Engineering Education. Additionally, the chapter delineates the process by which engineering education studies have been accorded the status of an independent academic discipline.

Keywords Engineering education research · Fields of research in engineering education · Independent second-level discipline

Engineering education research has always been an important part of the history of higher education research in China. However, in recent years, as an independent field of study, engineering education has received tremendous and rapid growth in social attention. Compared to the 1990s, when engineering education was only a small branch in the overall landscape of higher education research, today, engineering education research has taken up a prominent place in Chinese higher education research, especially since the beginning of the twenty-first century, representing approximately half of the total research output. Not only do research institutions closely related to engineering education support and strengthen engineering education as always, but also other research institutions, institutions of higher learning, and government departments have established various organizations and supported scholars to carry out engineering education research. Since the official launch of the new engineering education in 2017, the Ministry of Education has entrusted university scholars with various commissioned projects each year and provided them with significant funding. In 2022, the Ministry of Education even granted engineering education an independent disciplinary status, which put engineering education on an equal level with higher education, which was once considered higher in status than engineering education.

7.1 A Special Engineering Research Guideline

In fact, the undertaking of a large number of large-scale engineering education research is related to the launch of NEE in 2017. In the Beijing Compass, one of the three important documents of NEE, it clearly states that colleges and universities should assess the current situation of engineering and engineering education development, predict trends in advance, actively adapt and respond, give full play to the initiative of grassroots-level institutions and scholars, and explore the new concept of practice-oriented engineering education. The Beijing Compass specifically calls for scholars across the country to study key areas of engineering education, such as the new structure of disciplines and specialties, the new mode of talent training, the new quality of education and teaching, and the new system of classified development. The Beijing Compass has given rise to a new policy document, NEE Research and Practice Guideline, that provides a comprehensive framework and specific guidance for universities and colleges to implement NEE research projects in their engineering programs. It covers various aspects of the project, including the objectives, principles, implementation strategies, and evaluation mechanisms, and it identifies 24 fields in which engineering education research should be carried out.

The first research field concerns basic issues and problems in engineering education. Its goal is to analyze the opportunities and challenges faced by the development of new engineering courses, reveal the connotation, characteristics, laws, and development trends of new engineering, and refine the core objectives of training new engineering talents from a national, global, and future perspective. It also aims to provide new ideas and guidance for the development of engineering education in China. Relevant contents include the basic characteristics of the new industrial revolution and its impact on engineering education, the connotation, characteristics, regularities, and development trend of engineering technologies, the core objectives of training new engineering talents in the new era, the relationship between new engineering and traditional engineering and applied science, the main scope and disciplinary standards of new engineering, the opportunities and challenges faced by NEE, the paradigm shift of engineering education, the national responsibility, global obligation, and future mission that China's engineering education must undertake, and how to implement the national development concept of "innovation, coordination, green, open, and sharing." Other topics include new ideas for the reform and development of engineering education in China, key points, difficulties, and main tasks of implementing NEE in different types of colleges and universities.

The second field is about investigating and analyzing the demand for engineering talents in the new economy. Its goal is to conduct large-scale industrial enterprise research in different industries and regions around new technologies, new industries, new formats, and new models, providing basis and guidance for engineering specialty setting and personnel training in colleges and universities. Research contents include the demand and trend of engineering talents for new technologies such as big data, cloud computing, the Internet of Things, artificial intelligence, virtual reality, genetic engineering, nuclear technology and intelligent manufacturing, integrated circuit, aerospace and ocean, biomedicine, new materials, new energy and other new industries, providing data support, suggestions, and development ideas for the setting and development of new engineering majors, professional structure adjustment, and talent demand analysis.

The third field is the comparative research and analysis of international engineering education reforms. Its goal is to summarize and analyze from an international perspective and provide experience and reference for China's engineering education reform. Specific contents include analyzing the history and experience of engineering education reform in major developed countries, summarizing the law of interaction between higher education and previous industrial revolutions, and the trend of engineering education reform since the third industrial revolution. The content also includes a comparative analysis of various countries from the dimensions of background, policy, system and mechanism of engineering education development, personnel training mode, discipline and professional courses and teaching teachers, team building evaluation system, etc., so as to summarize the law of international engineering education development and the experience that can be used for reference.

The fourth field focuses on analyzing the reform and experience of engineering education in China. Its contents include an in-depth investigation into the implementation of China's engineering education reforms since the reform and opening up. This includes various state-level projects such as the education and training plan for outstanding engineers, the professional certification of new specialty development in strategic emerging industries, and the implementation and practice of the CDIO model. The research in this field aims to comprehensively summarize the development and experience of China's engineering education reform, reveal the laws of China's engineering education development, analyze the problems faced by China's engineering education currently, and based on the development trend, study and propose measures for the development of new engineering education. Relevant policies and supporting measures will also be suggested.

The fifth field aims to explore the practices of the transformation and upgrading path of engineering specialties that face the new economy. The goal is to meet the needs of transforming and upgrading traditional industries and cultivating and expanding emerging industries. It promotes the deep integration of knowledge, ability, quality, and requirements of high-tech and engineering majors, and explores the implementation path of transformation and upgrading of engineering majors. The specific content includes research and analysis of the new economy, which puts forward new requirements for the training of traditional engineering professionals. The curriculum system and teaching content will be updated, and the research will explore the ways and means of digital transformation of traditional engineering majors, as well as the ways and means of interdisciplinary and compound transformation of traditional engineering majors. It will also explore new technologies such as artificial intelligence, big data, cloud computing, and the Internet of Things.

Gradually, a new curriculum system based on the new direction and new field of transformation and upgrading of existing engineering majors will be formed.

The sixth field is the exploration and practice of the development of new engineering majors with multidisciplinary perspectives. Its goal is to promote interdisciplinary and cross-border integration according to the development trend of new technologies and new industries, to promote cross-integration between engineering majors and other disciplines, and to cultivate and build new engineering majors. The topic includes exploring and setting up new engineering majors for new technologies, new industries, and future technologies. It also involves formulating training programs for various emerging engineering majors characterized by interdisciplinary intersection, reorganizing and optimizing the curriculum system and teaching content of new engineering majors covering the basic knowledge of various disciplines, building a practical and innovative education and teaching system of new engineering majors, and studying the new engineering majors with interdisciplinary intersection, the requirements for teachers and the ways to achieve them.

The seventh field is to explore the development of new engineering majors derived from pure science. Its purpose is to explore the application of science in the forefront of technology with the goal of leading future technology and industry. It aims to promote the application of science and extend it to engineering, promote the interdisciplinary development of science, engineering, and other disciplines, and breed new engineering majors. Specific content includes focusing on national strategy and future industrial needs, exploring the environmental conditions and paths for applied science to produce new technologies, studying the cross-integration between science and engineering, promoting the cross-integration between applied science and fields such as environmental medicine, materials, energy, and artificial intelligence, and cultivating new engineering fields and related majors.

The eighth field is to study and explore the setting of engineering specialties and its dynamic adjustment mechanism. Its goal is to establish a professional setting and dynamic adjustment mechanism to meet the needs of industrial development and enhance the adaptability and supporting ability of engineering personnel training for national strategy and economic development. The specific contents include studying the characteristics and laws of industrial development, grasping the dynamic changes of industrial development and requirements for engineering talents, studying the specialty setting and dynamic adjustment mechanism and the adaptability between specialty and regional economic development, and putting forward operational suggestions and schemes for establishing specialty setting and its dynamic adjustment mechanism.

The ninth field is to study the mode reform and practice of new engineering education and multi-party collaborative education. Its goal is to further promote the open education ecosystem, establish a multi-agent collaborative education mode with the government, universities, and enterprises having a synergistic effect. Its contents include how to strive for various social resources, attract multiple parties to participate in the development of new engineering, innovate the multi-party collaborative education mode between universities and domestic and foreign industry enterprises, scientific research institutes, other universities, and local governments, and build a talent training community with complementary advantages, joint development of projects, sharing of achievements, and win–win benefits. In combination with the development trend of the new economy and the industrial demand, it aims to build a new engineering talent collaborative training mode with multi-agent participation and integration of production, education, and research. It also aims to promote the innovation of university organizations and explore the development of new organizational models such as industrial colleges with the participation of both inside and outside the university.

The tenth field is to study the training mode of engineering talents with interdisciplinary integration with the purpose of breaking the boundaries of inherent disciplines and forming an engineering talent training model that reflects the characteristics of interdisciplinary integration. Specific contents include how to optimize the organizational model of universities and colleges, establish a new type of interdisciplinary integration institution, and provide organizational guarantee for cross-department, cross-discipline, and cross-specialty training of new engineering talents. It also aims to reform the curriculum system, open interdisciplinary courses, and explore courses and teaching models for complex engineering problems. It also aims to establish interdisciplinary teaching teams, interdisciplinary project platforms, and promote interdisciplinary cooperative learning. Furthermore, it will study and formulate the evaluation criteria and assessment methods for the achievement of interdisciplinary integration capabilities and establish a quality monitoring system. Finally, it will carry out the construction of professional clusters that reflect the advantages and characteristics of the universities.

The 11th field is to explore and cultivate the innovative and entrepreneurial abilities of new engineering talents. Its goal is to improve the innovation and entrepreneurship education system for engineering talents and enhance their abilities in these areas, as well as the corresponding talent training patterns. Specific research contents include how to improve the course system and management system of innovation and entrepreneurship education on campus, strengthen the general education of innovation and entrepreneurship, and actively explore the setting of cutting-edge courses, comprehensive courses, problem-oriented courses, and interdisciplinary research courses. It also includes strengthening the innovation and entrepreneurship orientation of graduation design, exploring diversified training modes such as "engineering+" entrepreneurship double degree, major and minor system, and building a university innovation and entrepreneurship platform based on the advantages and characteristics of engineering to guide and encourage students to actively participate in innovation activities and entrepreneurship practice.

The 12th field is to explore the training mode of personalized talents in new engineering. Its goal is to implement the student-centered concept, explore ways to meet the individual needs of students, and form a learner-centered engineering education model. Specific contents include research and analysis of the thinking mode, behavior mode, learning objectives, and methods of college students in the Internet era, as well as summarizing the successful experience of personalized talent training in domestic and foreign universities. It also includes providing rich and varied courses and teaching resources, encouraging students to plan their career development independently under the guidance of their tutors, and allowing students to choose majors and freely combine courses. Furthermore, it includes summarizing the successful experience of the existing "Top Plan" to cultivate innovative students, exploring the personalized talent training mode of new engineering, and fully displaying the talent and specialty of students. Lastly, it includes studying and formulating the self-designed training program and the standards and procedures of the self-created specialty, establishing the necessary supporting conditions, and improving the personalized evaluation of talent training quality to promote the continuous improvement of the curriculum system and training programs.

The 13th field is to study the training mode of high-level talents in new engineering education. Its goal is to address the needs of high-level engineering and technological talents in the industry, and to explore the knowledge structure, curriculum system, training mode, and supporting system. Specific contents involve an in-depth analysis of the learning objectives, curriculum system, and learning outcome evaluation of learning at undergraduate, master, and doctoral levels. The focus is put on exploring the effective connection of core knowledge, ability, and quality requirements in different stages of related majors, and to create a multi-channel student development path. Engineering programs' course selection system is expected to be improved, including the credit confirmation mechanism for different majors, so will be the implementation of students' independent selection and student diversion programs based on corresponding admission conditions. Overall, this field aims to provide a comprehensive understanding of the curriculum system and training mode for high-level engineering and technological talents. It will explore effective approaches to develop and support students in different stages of their studies, ultimately facilitating the development of a skilled workforce for the industry.

The 14th field is dedicated to the development of quality standards for the training of emerging engineering professionals. Its goal is to customize the content of quality standards to meet the needs of emerging engineering professionals. The contents include exploring the forefront of international engineering education reform and development, assessing new trends in engineering education in developed countries, and working towards the collaboration between universities and industries with the goal of leading the world in facing the future. Both universities and enterprises are expected to jointly study and propose quality standards for the training of emerging engineering professionals. This includes the basic requirements of training specifications, curriculum system, teaching norms, teaching staff, and more. These standards serve as the basis for complying with the teaching quality evaluation of specialty setting and specialty construction, and for updating and improving training.

The 15th field aims to construct a new basic curriculum system for engineering education, along with a corresponding general education curriculum system. The goal is to integrate, optimize, and reorganize the basic curriculum system for engineering majors in accordance with the training requirements of new engineering talent, with the aim of improving students' learning efficiency and effectiveness. Additionally, the field explores how to cultivate critical thinking, design thinking, engineering thinking, digital thinking, engineering management thinking, engineering ethics, cross-cultural communication literacy, and other important skills among engineering students. The field also examines the digital thinking and abilities that new engineering talent should possess, in order to carry out the reform and practice of computer general education curriculum systems for non-computer engineering majors.

The 16th field aims to research how to build a practical platform for the engineering practice education system for new engineering education. Its contents include focusing on cultivating engineering students' practical skills, setting up curriculums, arranging training and practice, investing in funding, establishing institutional mechanisms, obtaining employer feedback, and other key links to carry out investigations. This field is to deeply analyze the current situation and problems related to engineering practice ability among engineering students in China. It promotes the design evaluation system for engineering students' practical training based on results orientation, guides reform practices, selects manufacturing and high-tech enterprises nationwide, establishes practice bases for engineering college students, forms a long-term and stable cooperative relationship between schools and enterprises, and puts forward relevant countermeasures and suggestions from the dimensions of supporting schools with policies, institutional mechanisms, and deep participation of enterprises.

The 17th field aims to explore the incentive mechanism for faculty development and evaluation in the construction of new engineering. Its contents include combining the characteristics of different types of universities and disciplines, strengthening teachers' engineering background and practice ability, putting forward clear requirements for teachers' industrial experience, actively creating conditions, exploring the path of teaching staff construction that matches with new engineering courses, and formulating and implementing standards for the classification evaluation of instructors.

7.2 Reframing Engineering Education as an Independent Second-Level Discipline in China's Higher Education Landscape

Within China's higher education landscape, the granting of independent status to a discipline signifies its increasing recognition and importance. China's catalogue of higher education programs comprises numerous first-level and second-level disciplines. First-level disciplines, such as economics, sociology, and pedagogy, are well-known to the general public. Second-level disciplines represent further subdivisions within first-level disciplines. For instance, within the first-level discipline of economics, applied economics, national economics, macroeconomics, and microeconomics are second-level disciplines. Likewise, under the first-level discipline of pedagogy, higher education, comparative education, special education, and educational psychology are examples of second-level disciplines.

Historically, engineering education has been considered a lower-level branch of higher education, lacking the recognition of an independent second-level discipline.

Although widely acknowledged, this classification has led to insufficient attention being paid to engineering education as a distinct area of scholarship when policymakers formulate crucial educational policies.

In October 2022, this situation underwent a significant change. The Academic Degree Office of the State Council officially designated engineering pedagogy as an independent second-level discipline, on par with higher education, comparative education, and other disciplines. This decision granted engineering education an independent and substantively equivalent academic status. Ten universities in China were selected to pioneer the implementation of engineering education as a second-level discipline. These institutions include Tsinghua University, Beijing University of Aeronautics and Astronautics, Zhejiang University, Beijing Institute of Technology, Tianjin University, Nanjing Normal University, Dalian University of Technology, South China Normal University, Northwestern Polytechnical University, and Huazhong University of Science and Technology.

The ten universities possess both traditional engineering education strengths and comprehensive academic offerings. For example, while Nanjing Normal University and South China Normal University primarily emphasize the quality of humanities and social sciences, their respective teams have been deeply engaged in the field of engineering education for many years. This development in China's higher education landscape marks a critical step in acknowledging and supporting the vital role of engineering education.

The newly-built second-level discipline is also of great significance to the universities where they are located. Not only have these universities obtained the qualifications of independent enrollment and running schools, but they will also rely on them to obtain a large number of funds for research in the field.

Chapter 8
Impact, Challenges and Prospects of the Engineering Education Reform

Abstract This chapter comprehensively elucidates the impact of a series of measures taken in the recent years concerning China's engineering education reform, the challenges faced, and its future prospects. It delves into the structural and cultural transformations of Chinese engineering education under the guidance of the new paradigm of engineering education. These influences and changes manifest both at the macroscopic and microscopic levels, with the challenges predominantly occurring at the microscopic scale. The chapter also highlights the specific conditions required for the further high-quality advancement of engineering education reform in the future.

Keywords Structural changes · Cultural changes · Challenges and prospects

In recent years, the reform of engineering education in China has undergone significant advancements characterized by seriousness, systematicity, and comprehensiveness. This reform has been undertaken under the direct guidance and promotion of the Ministry of Education, particularly the Department of Higher Education. Higher education institutions across the nation have actively responded to the development requirements of NEE, which also facilitate the reform of engineering education by tailoring it to the unique conditions and characteristics of each institution. The series of measures implemented has resulted in profound transformations in the structure and teaching culture of engineering education at university level. Structural changes include updating existing professional content and establishing emerging majors to recombine corresponding resources to promote interdisciplinary nature of the profession, and updating the main structure of engineering education implementation through industry university cooperation and collaborative education. The corresponding cultural changes are manifest but not limited in the fact that optimizing and updating the teaching methods and curriculum content has become a critical consensus for colleges and universities that offer engineering programs, the further improved state of engineering education in the higher education landscape, the emphasis of interdisciplinarity and cross-disciplinary collaboration in offering engineering programs, the underscoring of contextualization of knowledge and combining theoretical knowledge with praxis in engineering education, the advancement of a synergistic approach to engineering talent cultivation based upon teaching-focused university industry collaboration, etc.

8.1   The Structural Changes of Engineering Education at University Level Under NEE

According to the statistics of Ministry of Education in 2022, China has established the world's largest higher education system, with a total enrollment of over 44.3 million students. The gross enrollment rate in higher education has increased from 30% in 2012 to 57.8% in 2021, a significant increase of 27.8 percentage points. This achievement represents a historic leap, marking the transition of higher education into a universally recognized stage of popularization worldwide. The population receiving higher education in China has reached 240 million, with an average education duration of 13.8 years for the new labor force. The structure of the labor force has undergone significant changes, leading to a steady improvement in the overall quality of the nation. Higher education has continuously innovated in terms of nurturing methods, educational models, management systems, and guarantee mechanisms.

Among these innovations, the development of engineering education has made a significant contribution to the overall advancement of higher education. In order to proactively respond to the new round of technological revolution and industrial transformation, promote reform and innovation in engineering education, and accelerate the cultivation of outstanding engineers capable of meeting the future social development needs, the Ministry of Education launched the New Engineering Education Initiative in 2017. This initiative has become one of the most influential and wide-ranging reforms in recent years, aimed at cultivating talents in higher education with a focus on the future, proactive adaptation, and maximum impact. Today, the New Engineering Education Initiative has achieved remarkable results, serving as a powerful driving force in building a distinctive, world-class engineering education system in China and accelerating the country's transformation from being a major engineering education nation to becoming a powerhouse in engineering education. The New Engineering Education Initiative represents a new paradigm for teaching reform in higher engineering education, with its core goal being the enhancement of the quality of talent cultivation and active response to the urgent demand for outstanding engineers in the new era. After five years of exploration by researchers and practitioners in the field of New Engineering Education, a number of exemplary models for the construction of the New Engineering Education Initiative have been developed, and three consensuses on talent cultivation have been reached, namely, emphasizing interdisciplinary integration in-depth, focusing on the integration of general and specialized curriculum systems, and prioritizing deep integration of industry and education. It is reported that the deepening of the New

Engineering Education Initiative has led to comprehensive innovation in the organizational model, theoretical research, content and methods, and practical systems of higher engineering education. A total of 1457 New Engineering Education projects have been implemented, exploring the construction of an educational mechanism that integrates multiple elements of industry, academia, research, and application and promotes multi-stakeholder collaboration. Over 1100 undergraduate institutions have partnered with nearly 800 enterprises, resulting in 37,000 collaborative projects being approved, with the enterprises providing approximately 11.2 billion yuan in funding and support for hardware and software. By focusing on the innovation of the organizational model of colleges and departments, the construction of 28 exemplary microelectronics colleges, 11 top-level network security colleges, 50 modern industry colleges, 33 exemplary software colleges, and the initial establishment of 12 future technology colleges has been promoted, driving deep-level transformations in engineering education. In terms of improving the system for awarding degrees and the catalog of majors in universities, engineering education has actively integrated into national strategies and industry developments, adjusted and optimized the layout of disciplinary majors to align with the new development landscape, and promoted interdisciplinary integration to address real-world issues. Since 2012, a total of 265 new majors have been included in the undergraduate catalog, bringing the current total to 771 majors. There have been 17,000 new undergraduate programs established, while 10,000 programs have been discontinued or suspended, demonstrating a noticeable increase in the adaptability of talent cultivation to new technologies (Department of Higher Education, 2022).

Structural changes are not only reflected in the overall trend of engineering programs across institutions, but also mirrored in micro-level course development and program adjustment for individual higher education institutions. For instance, The University of Electronic Science and Technology (UESTC) seizes the opportunity to adjust its professional structure and establish a new paradigm of rapid response and quality improvement. Focusing on the "high-precision, cutting-edge, and scarce" fields of engineering, UESTC aims to establish a dynamic mechanism for professional adjustment that encompasses the enrollment and cultivation processes, with the goal of enhancing its ability to support and contribute to economic and social development. It specifically targets "bottleneck" technological areas such as core electronic materials and components, high-end integrated circuit chip design and manufacturing, and network security. UESTC has introduced majors such as integrated circuit design, integrated systems, and network cybersecurity, while also establishing a national innovation platform for the integration of industry and education in the field of integrated circuits, as well as a national exemplary software college. In response to the trends of "intelligence+" and "smart+", UESTC is also developing eight emerging interdisciplinary majors that integrate artificial intelligence, robotics engineering, intelligent grid information engineering, and internet finance. Adhering to the principles of grouping admissions, cultivating students in groups, and managing programs in groups, UESTC has established a common module for core courses in high-standard, high-demand, and high-value-added majors. It places particular emphasis on strengthening the foundation of mathematics and natural science courses and aims to cultivate students' deductive reasoning and logical proof abilities. UESTC actively promotes the national engineering education program accreditation for its engineering majors. Currently, 32 majors have been selected as national first-class professional construction sites, ensuring comprehensive coverage of electronic information and computer majors at the university. UESTC has made breakthroughs in disciplinary majors, exploring new pathways for cross-school interdisciplinary education. It has created a cross-school interdisciplinary training program that combines "New Engineering" and "New Business" and has jointly developed a bachelor's degree program in "Intelligence + Blockchain Finance" in collaboration with other universities. Through integrated programs, cohesive curriculum systems, and comprehensive projects, UESTC aims to cultivate talents with expertise in both computer science and finance. Furthermore, UESTC collaborates with relevant universities and enterprises to establish an interactive experimental art program that integrates "New Engineering" and "New Art." By organizing challenging experimental art projects, art exhibitions, and other activities within the university, between universities, and in collaboration with enterprises, UESTC aims to cultivate engineers with artistic literacy and practical skills. By integrating high-quality teaching resources from six colleges, four major disciplines, and six majors, UESTC has created an experimental class for dual degrees in "Internet+" education. Through customized training programs, interdisciplinary courses, and innovation and entrepreneurship projects that integrate industry and education, UESTC strives to cultivate innovative talents who can contribute to the country's digital strategy. Additionally, UESTC has established a joint training program for "Management-Electronic Engineering," promoting the organic integration of capabilities, qualities, innovation, and entrepreneurship. The program aims to cultivate talents with both economic management capabilities and information technology literacy (MOE, 2022).

Structural adjustments of this nature are prevalent in various higher education institutions. Failure to undertake such adjustments may result in the inability to meet the scrutiny of pertinent authoritative bodies, leading to missed opportunities for further support.

8.2   The Changes in the Culture for Teaching and Learning Engineering Subjects at University Level

One of the most significant aspects of any higher education reform is the evolution of teaching and learning methods. Traditional pedagogical models are increasingly inadequate in catering to the dynamic needs and skills of 21st-century students. Students today require not only knowledge but also skills to analyze, create, and innovate. With technological advancements and the evolution of society, the demand for active learning strategies, digital literacy, and interdisciplinary knowledge has grown significantly (Fullan & Langworthy, 2014). The transition from passive to active learning methods is a crucial aspect of teaching and learning changes. Traditional lecture-based models are increasingly replaced by methods that encourage students to engage more actively in their learning. This can include problem-based learning, flipped classrooms, or collaborative learning experiences. Research has repeatedly shown that active learning strategies improve student engagement, retention, and outcomes (Freeman et al., 2014). In this context, pedagogical innovation becomes a key element of educational reform, and changes in teaching and learning are not only desirable but essential in higher education reform. To create a future-proof education system that prepares students for the complex and rapidly evolving world, pedagogical innovation is of paramount importance.

Over the past decade, China's engineering education has undergone significant transformations, underpinning the country's ambition to become a global powerhouse in innovation and technological advancement. These sweeping reforms have been strategically implemented, impacting not just the course content, but also shifting the broader teaching and learning culture of engineering subjects at the university level. China's Ministry of Education has spearheaded these reforms, heavily investing in the improvement of engineering education to help China transition from a manufacturing-based to an innovation-driven economy. In 2017, the Ministry of Education launched the "New Engineering" initiative, aiming to modernize the traditional engineering education by promoting interdisciplinary knowledge and encouraging the adoption of emerging technologies.

The reform focused on integrating disciplines and enhancing the digital competency of engineering students. The program's emphasis on 'fusing' and 'emerging' reflects the importance of cross-disciplinary collaboration and the need for students to be well-versed in emerging technologies such as AI, IoT, and data science. This shift required universities to update their curriculums and teaching practices, fostering a culture of interdisciplinary collaboration and innovative thinking among students and faculty alike. The NEE initiative also fostered a change in teaching culture, with universities encouraged to adopt a project-based and problem-solving approach. This approach was rooted in experiential learning, with students required to apply their theoretical knowledge to practical situations, developing their problem-solving and critical thinking skills. Instead of the traditional lecture-based teaching, this method was more hands-on and student-centric, fostering a more active and engaging learning environment.

Moreover, the reforms emphasized cultivating students' entrepreneurial and innovative capabilities, recognizing the need for engineers to be not just problem-solvers, but also creators and innovators. This was in line with China's ambition to become a global innovation hub and was reflected in the introduction of entrepreneurship courses in engineering programs, as well as the establishment of innovation and entrepreneurship centers within universities. The impact of these reforms on university-level engineering education in China has been substantial. They have fostered a new culture of teaching and learning, characterized by interdisciplinarity, innovation, and hands-on problem solving. These changes reflect a broader global shift in engineering education, with many countries recognizing the need to prepare students for the increasingly complex and fast-paced technological landscape.

8.3   Remaining Challenges

Despite the remarkable progress of China's engineering education advancement, some challenges remain for the NEE initiatives and the overall engineering education plans to achieve original goals. The hierarchical architecture of Chinese universities, in regard to both the quality of education and the level of institutional support provided, coupled with the intricacy of systemic modifications requisite for efficacious reforms, indeed present considerable challenges. Of paramount concern, however, are the limitations that educators in the field of engineering encounter within these higher education frameworks. These factors collectively constitute potential impediments to the successful realization of these much-needed reforms.

The higher education sector in China has experienced considerable expansion since the onset of the twenty-first century. However, an observable stratification among higher education institutions regarding the quality of education persists. As of 2016, out of a total of 2596 higher learning institutions nationwide, a significant majority, amounting to 1359, are classified as higher vocational colleges. The 'New Engineering Education' initiative is primarily directed at universities, excluding these higher vocational colleges. Among the 1237 universities distributed throughout the country, a mere 39 have been selected for the prestigious '985 Project,' and only 100 have been incorporated into the '211 Project.' This represents a proportion of only 3.15% and 8.08% respectively, illuminating the stark disparity within the system (Zhuang & Xu, 2018).

Drawing on a range of scholarly literature addressing the quality hierarchy and societal acknowledgment of Chinese universities, Fig. 8.1 offers a general depiction of the pyramid-like structure within China's higher education institutions. The pinnacle of the pyramid represents those universities incorporated in the '985 Project', encompassing 39 institutions in total. The subsequent tier includes non-985 institutions that have been incorporated into the '211 Project'. Both the '985 Project' and '211 Project' were initiated by the Ministry of Education towards the close of the twentieth century, and the majority of the selected institutions in these projects are comprehensive universities with a few exceptions. Below these top tiers, institutions specializing in specific industry-related disciplines, such as telecommunications, finance, economics, and oil, occupy the third tier. However, they lag behind the aforementioned comprehensive universities in terms of societal recognition and governmental support. The bottom tier comprises the remaining institutions nationwide, making up the largest segment of the entire higher education sector, inclusive of vocational institutions.

This pyramid structure signifies stark disparities among universities at various levels. The financial burden of supporting the substantial number of institutions, particularly those in the bottom two tiers, largely falls upon provincial governmental bodies, while only a small proportion of top-tier institutions receive direct state support. These funding differences not only reveal the disparate levels of official expectations for these institutions but also correlate with the degree of societal recognition. This recognition impacts not only the visible funds an institution

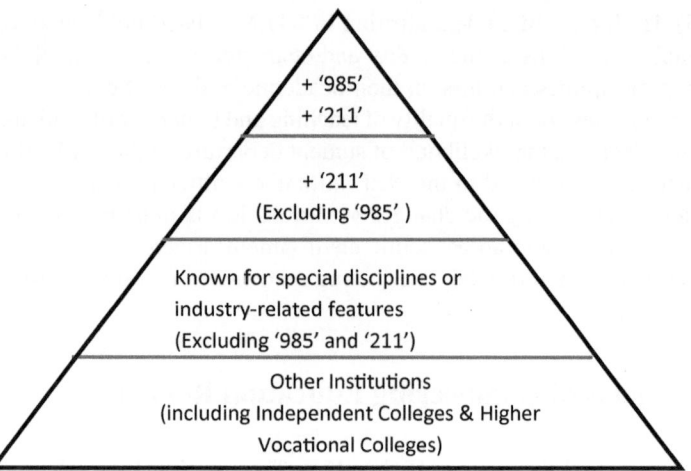

Fig. 8.1 Pyramid for China's higher education institutions. *Source* Zhuang and Xu (2018)

may obtain but also the quality of enrolled high school graduates. Furthermore, it affects the volume of research grants that institutions can obtain from the government. For instance, in 2017, Tsinghua University unsurprisingly topped the list of total research grants in China, obtaining around 1.6 billion RMB, followed by other first-tier institutions such as Shanghai Jiaotong University, Nanjing University, and Zhejiang University. Comparatively, second-tier universities selected for the '211 Project' but not the '985 Project' receive grants amounting to roughly one-sixth of their '985' counterparts.

At the same time, the existing higher education system in mainland China is not devoid of constraints in terms of facilitating necessary systemic changes, particularly concerning the inefficiencies generated by bureaucratization. Some have identified five major stakeholders in China's engineering education sector: students, teachers, universities, employers, and governments (Jia & Xiao, 2009). Among these, excessive governmental influence and insufficient employer influence have led to the bureaucratization of the higher education system, detracting from the core focus of teaching the next generation of engineers. Some others' analysis of the academic identities of those working in Chinese universities renowned for their engineering programs has resulted in the categorization of Chinese academics into six distinct types: the managerial advocate, the academic chameleon, the knowledge worker, the stressed faculty, the resolute pilgrim, and the careless outsider. These categorizations reveal a competitive and stressful environment for Chinese university instructors, suggesting that genuine curiosity and intellectual passion are often eclipsed by external bureaucratic factors (Huang, 2020; Huang & Xu, 2019).

Nevertheless, numerous international studies confirm that academic quality, an engaging learning environment, and highly competent instructors with deep content knowledge in engineering subjects and proficient pedagogical skills are crucial factors for ensuring excellence in STEM learning (Martinez Ortiz & Sriraman, 2015;

Xu, 2015). High rates of student attrition in STEM-related fields tend to rely more on students' perceptions of the quality and character of science, mathematics, and engineering disciplines and less on their academic abilities (Seymour et al., 1997). Positive perceptions about the quality of teaching and their overall academic program significantly decrease the likelihood of student departure. If the cultivation of excellent engineers, as expressed in the NEE initiative's official documents, is to be the goal, systemic and synergetic changes at various levels must be effectively implemented. This will enable an academic environment driven by genuine intellectual curiosity and passion, ultimately facilitating innovative creations across disciplines.

8.4   Prospects of Engineering Education Reform

The purpose behind the launch of the 'New Engineering Education' initiative is twofold. Primarily, at a national level, the objective is to guarantee the sustainable production of engineering talents (human capital) by improving the quality of engineering education. This is aimed to support China's economic transition and ascendancy. Secondary to this, from the perspective of higher education itself, China aims to shift from the 'periphery' to the 'center' in the global higher education landscape. The justification for this initiative can be interpreted through the lens of human capital theory, development theory, and the center-periphery model as discussed in prior sections.

In the last decade, China has achieved significant advancements in its engineering education sector, thereby paving the way for the 'New Engineering Education' reform initiative. The observable goals indicated by official documents appear to culminate in the realization of the 9 'bunches' within a decade or so. However, these documents do not explicitly designate which types of institutions should develop which 'bunches', considering the drastically disparate development and the vast number of higher education institutions throughout the country. This poses both pros and cons for individual institutions. The advantage is that it provides institutions the flexibility to grow according to their unique circumstances and available resources. On the downside, many institutions may struggle to obtain resources and form beneficial partnerships without a clear official directive in such a centralized system. Furthermore, the structure of university-industry partnerships in China is not as developed as in advanced nations such as Germany and the United States. Against this background, the ability of institutional leaders plays a significant role in securing resources and forging partnerships.

The success of the NEE initiative can only be assessed over a medium to long-term timescale, such as a decade or longer, given the timeframe required for talent cultivation. The quality of the 'human capital' and the effectiveness of student development in terms of knowledge, skills, and competences will be evaluated by various stakeholders, including academics, the industry, and society at large. An alarming trend at present is the apparent devaluation of teaching in many universities due to the emphasis placed on research and publication in top-tier journals. An insightful quote by one of Tsinghua University's most esteemed presidents, Mei Yi-chi in 1931, rings true: 'What makes a good university is not the number of buildings it can have, but the number of great scholars, professors, and masters.' While President Mei's tenure occurred during the Republic of China period, his sentiment remains relevant in China's higher education history and continues to influence several generations of scholars across the Taiwan Strait. In China, a nation that endured prolonged periods of impoverishment and foreign invasion, the significance of this sentiment and the impact of 'great scholars, professors, and masters' was more profoundly reflected in educating students—the future nation-savers—than in conducting pure research. Applying this concept to the current 'New Engineering Education' initiative, it is suggested that offering high-quality and dedicated teaching to equip students with a strong engineering knowledge base and skills to tackle challenges should be considered equally, if not more important, than publishing SCI or SSCI journal articles for university instructors.

As pointed out by Borrego and Henderson (2014), achieving comprehensive reform in higher engineering education requires multiple levels of synergistic and collaborative change. China faces these challenges as it undertakes consistent and effective reform measures to ensure individuals are genuinely motivated and equipped to improve and progress. We conclude that in the coming years, while top Chinese universities are likely to continue ascending the global ranking ladder and lead the Chinese higher education sector, including the engineering education sector, there will be widening disparities and stratifications among different institutional tiers. Additionally, despite an expected rise in the overall quality of graduates with a bachelor's degree in engineering, the pace may not.

References

Borrego, M., & Henderson, C. (2014). Increasing the use of evidence-based teaching in STEM Higher education: A comparison of eight change strategies: Increasing evidence-based teaching in STEM education. *Journal of Engineering Education, 103*(2), 220–252. https://doi.org/10.1002/jee.20040

Department of Higher Education, F. of E. (2022). *Historical achievement and situational change: Achievements of higher education reform in the last decade.* Retrieved from http://www.moe.gov.cn/fbh/live/2022/54453/sfcl/202205/t20220517_627973.html

Freeman, S., Eddy, S. L., McDonough, M., Smith, M. K., Okoroafor, N., Jordt, H., & Wenderoth, M. P. (2014). Active learning increases student performance in science, engineering, and mathematics. *Proceedings of the National Academy of Sciences, 111*(23), 8410–8415.

Fullan, M., & Langworthy, M. (2014). *A rich seam: How new pedagogies find deep learning.*

Huang, Y.-T. (2020). Responding to the neoliberal and managerial changes: A generational perspective of Chinese academics. *Compare: A Journal of Comparative and International Education,* 1–19. https://doi.org/10.1080/03057925.2020.1716305

Huang, Y.-T., & Xu, J. (2019). Surviving the performance management of academic work: Evidence from young Chinese academics. *Higher Education Research & Development,* 1–15. https://doi.org/10.1080/07294360.2019.1685946

Jia, G., & Xiao, C. (2009). Research on five stakeholders & five relationships of higher engineering education in China. *International Journal of Modern Education and Computer Science, 1*(1), 60.

Martinez Ortiz, A., & Sriraman, V. (2015). Exploring faculty insights into why undergraduate college students leave STEM fields of study—A three-part organizational self-study. *American Journal of Engineering Education (AJEE), 6*(1), 43. https://doi.org/10.19030/ajee.v6i1.9251

MOE. (2022). *UESTC focuses on new engineering education development for better engineering education personnel training.* Retrieved from http://www.moe.gov.cn/jyb_xwfb/s6192/s133/s211/202211/t20221102_693071.html

Seymour, E., Hewitt, N. M., & Friend, C. M. (1997). *Talking about leaving: Why undergraduates leave the sciences* (Vol. 12). Westview press Boulder, CO.

Xu, Y. J. (2015). Attention to retention: Exploring and addressing the needs of college students in STEM majors. *Journal of Education and Training Studies, 4*(2). https://doi.org/10.11114/jets.v4i2.1147

Zhuang, T., & Xu, X. (2018). 'New engineering education' in Chinese higher education: Prospects and challenges. *Tuning Journal for Higher Education, 6*(1), 69–109.